TO THE FAR NORTH

A volume in the

NIU Series in Slavic, East European, and Eurasian Studies
Edited by Christine D. Worobec

For a list of books in the series, visit our website at cornellpress.cornell.edu.

TO THE FAR NORTH

DIARY OF A RUSSIAN
WORLD TRAVELER

IVAN NIKOLAEVICH AKIF'ËV

Translated by Andrew A. Gentes

NORTHEN ILLINOIS UNIVERSITY PRESS
AN IMPRINT OF
CORNELL UNIVERSITY PRESS
ITHACA AND LONDON

First published 2024 by Cornell University Press

Library of Congress Cataloging-in-Publication Data

Names: Akif'ev, Ivan Nikolaevich, 1872–1905, author. | Gentes, Andrew Armand, 1964– translator.
Title: To the far north : diary of a world traveler / Ivan Nikolaevich Akif'ëv; translated by Andrew A. Gentes.
Other titles: Na dalekiĭ siever. English
Description: Ithaca : Northen Illinois University Press, an imprint of Cornell University Press, 2024. | Series: NIU series in Slavic, East European, and Eurasian studies | Translation of: Na dalekiĭ siever, published: S. Peterburg : Tipo-lit. "Evgeniia Tile premn.", 1904. | Includes bibliographical references and index. | In English, translated from the Russian.
Identifiers: LCCN 2023040596 (print) | LCCN 2023040597 (ebook) | ISBN 9781501774607 (hardcover) | ISBN 9781501774614 (paperback) | ISBN 9781501774621 (epub) | ISBN 9781501774638 (pdf)
Subjects: LCSH: Akif'ev, Ivan Nikolaevich, 1872–1905—Travel. | Akif'ev, Ivan Nikolaevich, 1872–1905—Diaries. | Physicians—Travel. | Physicians—Russia—Diaries. | Voyages around the world. | Chukchi Peninsula (Russia)—Description and travel. | Sakhalin (Sakhalinskaia oblast', Russia)—Description and travel. | Alaska—Description and travel. | United States—Description and travel.
Classification: LCC G296.A55 A3 2024 (print) | LCC G296.A55 (ebook) | DDC 910.911/3 [B]—dc23/eng/20231011
LC record available at https://lccn.loc.gov/2023040596
LC ebook record available at https://lccn.loc.gov/2023040597

Contents

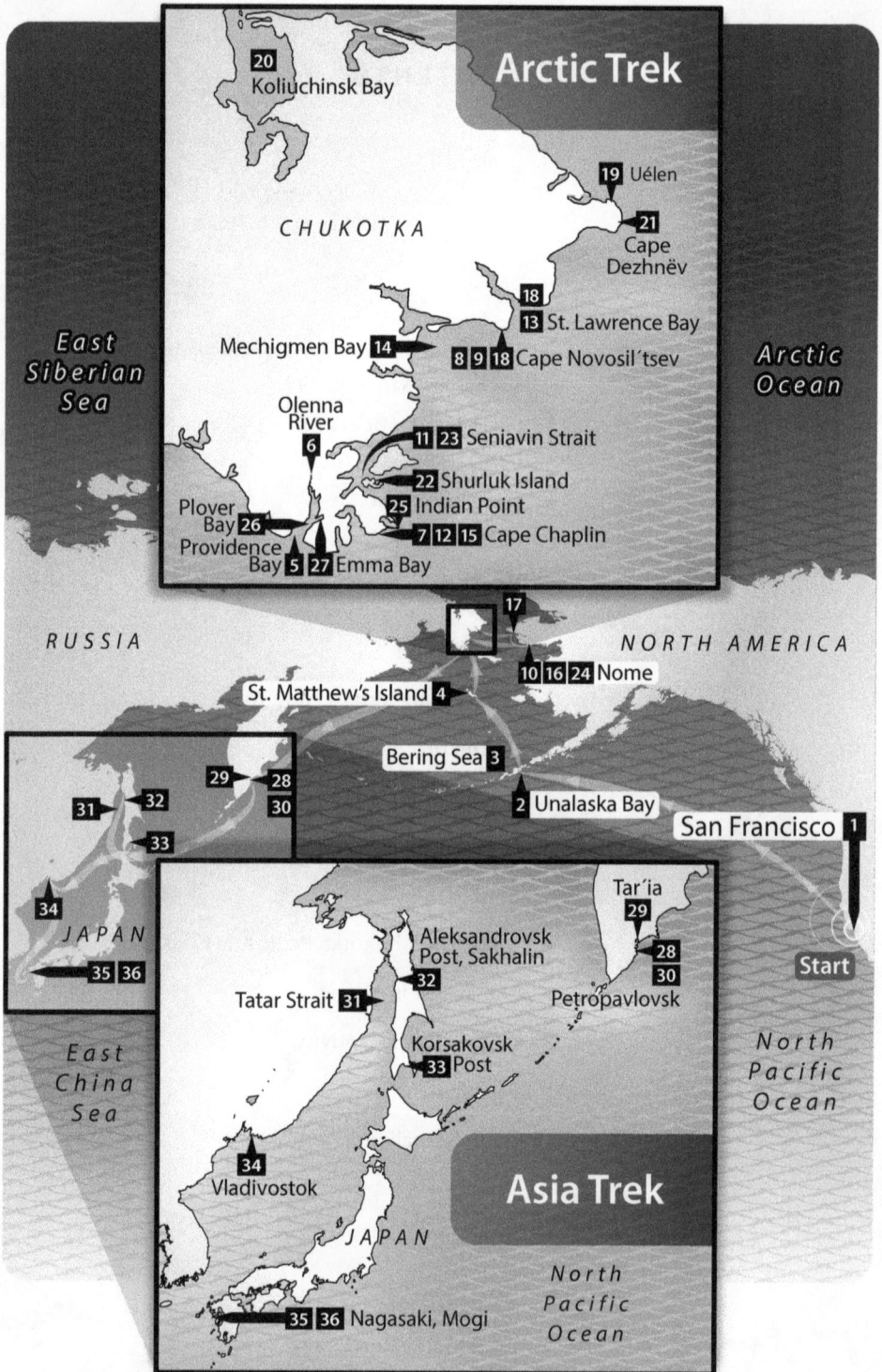

MAP 1. Map showing Akif'ëv's passage between North America and Russia

Chronology of Akif'ëv's Diary (Old Style/New Style Dates)

18 April/1 May 1900: Departure from St. Petersburg
20 April/3 May: Arrival in London
22 April/5 May: Departure aboard the *Compania*
29 April/12 May: Arrival in New York
6/19 May: Departure by railway from Washington, DC
9/22 May: Wyoming/Utah

Chronology (continued with reference to the map)

1. 27 May/9 June: Departure from San Francisco aboard the *Samoa*
2. 7/20 June: Arrival at Unalaska Bay
3. 9/22 June: In the Bering Sea
4. 10/23 June: Passing by St. Matthew's Island
5. 13/26 June: Arrival at Providence Bay, Chukotka Peninsula
6. 15/28 June: Prospecting up the Olenna River
7. 17/30 June: Visit to Unyyn village on Indian Point, Cape Chaplin
8. 18 June/1 July: Off Cape Novosil'tsev
9. 21 June/4 July: Departure from Chukotka for Nome
10. 23 June/6 July: First arrival at Nome
11. 27 June/10 July: Arrival at Seniavin Strait, Chukotka
12. 1/14 July: Arrival at and immediate departure from Cape Chaplin
13. 2/15 July: Arrival at St. Lawrence Bay
14. 9–11/22–24 July: Stuck in Mechigmen Bay
15. 13/26 July: Arrival at the Marich River, departure from Cape Chaplin
16. 15/28 July: Second arrival at Nome
17. 17/30 July: Departure for Cape York, Alaska
18. 18–20 July/1–3 August: Departure from Cape York, explorations in St. Lawrence Bay, Cape Novosil'tsev, Cape Litke (all on Chukotka)
19. 21 July/4 August: Arrival at village of Uélen, Cape Dezhnëv
20. 28 July/11 August: Departure from Koliuchinsk Bay

21. 29 July / 12 August: Return to Cape Dezhnëv
22. 31 July–5 August / 13–18 August: Explorations at Cape Novosil'tsev, Shurluk Island, Seniavin Strait
23. 6 / 19 August: Departure from Seniavin Strait and hijack to Nome
24. 7–16 / 20–29 August: Third arrival (and arrest) at Nome
25. 17 / 30 August: Arrival at Indian Point
26. 18 / 31 August: Arrival at Plover Bay
27. 26 August / 8 September: Departure from Emma Bay aboard the *Iakut*
28. 31 August / 13 September: Arrival at Petropavlovsk, Kamchatka
29. 8–9 / 21–22 September: Overnight excursion to Tar'ia
30. 11 / 24 September: Departure from Petropavlovsk aboard the *Peyang*
31. 15 / 28 September: Passage through the Tatar Strait
32. 16–23 / 29 September–6 October: Arrival and stay at Aleksandrovsk Post, Sakhalin
33. 25 September / 8 October: Arrival at Korsakovsk Post, and departure aboard the *Sungari*
34. 29 September / 12 October–8 / 21 October: Arrival and stay at Vladivostok, departure aboard the *Tartar*
35. 9 / 22 October: Arrival in Nagasaki, Japan
36. 29 October / 11 November: Departure from Nagasaki aboard the *Sydney* to begin the return to Russia

Dramatis Personae

The Russians

Akif'ëv, Ivan Nikolaevich—Physician and adventurer.

Beliaev, Vasilii—Russian American sailor.

Bogdanovich, Karl Ivanovich (Karol Bohdanowicz)—Polish Russian mining engineer and expedition leader. Bogdanovich was accompanied by his unnamed wife.

Lemashevskii—Reserve captain of the *Samoa*.

Makferson, Al'fred Markovich—Russian Englishman; jack-of-all-trades.

Miagkov, Aleksandr Gennad'evich—Mining specialist; Akif'ëv's travel companion; affectionately referred to as "Gennad'ich."

V–ii—Vladimir M. Vonliarliarskii, expedition organizer and investor.

Nikolai S.—Ossetian Terek Cossack; expedition assistant.

Petrov—Senior laborer.

Sanderson, James (Timofei)—Russian American sailor.

Shurduk, Grigorii—Russian American sailor.

Tret'iakov—Cossack commander.

The Anglos

Baker, Frederick W.—British investor; Vonliarliarskii's business partner; founder of the East Siberian Syndicate.

Bernstein—American purchasing commissioner.

Dowlen, John W.—American; Rickard's assistant engineer.

Evans, Joseph—Alaska customs agent.

Jahnsen, Edward—Norwegian American captain of the *Samoa*.

Jarvis, David H.—Alaska customs agent.

Kinney—American; the *Samoa*'s senior ship mechanic.

Landfield, J. B.—American friend of Rickard; mining engineer; Roberts's private secretary.

Picard—American; former taxidermist.

Reiner—American; the *Samoa*'s senior assistant captain.

Rickard, Forbes—Swiss-born resident of England and later, the United States; mining engineer.

Roberts, George D.—American millionaire; mining engineer consultant; Frederick Baker's business associate.

Scranton, Charles Sr.—American carpenter; mining engineer.

Scranton, Charles Jr.—Charles's adult son; former cannery worker; assistant engineer.

Stern, E. L.—Jewish German American purchasing agent with the rank of major (real or imagined).

The Stuarts—A mysterious couple never fully identified.

White, John A.—American; former diplomat and attorney.

Others

Chinese laborers
Chukchi

Guide to Measurements

arshin (arshiny, pl.)—A unit of length equal to 71 centimeters or 28 inches.

desiatina (desiatiny, pl.)—A unit of area equal to 1.09 hectares or 2.7 acres.

pood—Anglicization of *pud*, a unit of weight equal to 16.38 kilograms or 36 pounds.

sazhen´ (sazheni, pl.)—A unit of length equal to 2.13 meters or 6.99 feet.

vershok (vershki, pl.)—A unit of length equal to 4.45 centimeters or 1.75 inches.

verst—Anglicization of *versta*, a unit of distance equal to 1.06 kilometers or 0.66 miles.

zolotnik (zolotniki, pl.)—A unit of weight equal to 4.26 grams or 0.15 ounces.

NOTE ON THE TRANSLATION

Ivan Nikolaevich Akif'ëv's published diary appeared in two editions. The first was titled *Na dalekii sever za zolotom: iz dnevnika krugosvetnogo puteshestviia 1900 goda* (*To the Far North for Gold: Diary of a Global Circumnavigation of 1900*) and was published in St. Petersburg in 1902 by Evgenii Tile's publishing house. The house published a second edition in 1904, under the title *Na dalekii sever: iz dnevnika krugosvetnogo puteshestvennika* (St. Petersburg: Tipo-litografiia E. Tile preemn., 1904), which I have translated here as *To the Far North: Diary of a Russian World Traveler*. Their shared pagination suggests the two editions do not differ from each other. The present translation is based on this second edition because it was the one available to me.

Na dalekii sever was rather sloppily put together. This might be excused for what was, after all, a diary. But textual evidence indicates that certain entries were rewritten (if only to foreshadow later developments), and anyway the editing was poor for what was, after all, a commercial publication. I have corrected numerous misspellings and typographic errors as well as Akif'ëv's maddening inconsistency regarding proper nouns. He also had a habit of using semicolons in place of colons and was inconsistent in using spelled-out numbers and integers. But to reproduce his style and avoid reengineering his grammar, I have simply inserted colons and let the numbers appear as they do. Finally, I have rejiggered the beginnings and conclusions of his haphazard paragraphs to improve clarity and readability.

For some reason, Akif'ëv dated his diary entries inconsistently, sometimes using the Julian ("Old Style" [o.s.]) calendar, to which Russia adhered until 1917, and sometimes the Gregorian ("New Style" [n.s.]) calendar, which until 16 February 1900 (o.s.) was twelve days, and therefrom, thirteen days, ahead (all dates in this book's introduction and notes are in Old Style unless otherwise noted). To spare the reader unnecessary confusion, I have assigned both calendars' dates to each of Akif'ëv's entries and to the chronology corresponding to the map of Akif'ëv's journey.

Finally, Akif′ëv indicates that he and his traveling companion Miagkov took photographs throughout the entirety of their journey together. But the thirty-seven photos that appeared in his published diary for some reason cover only the Chukotka Expedition. A bound volume titled *Ekspeditsiia polkovnika V. M. Vonliarliarskago na Chukotskii poluostrov v 1900 godu* collates photos of Chukotka that were, however, *not* included in his book. Regardless, this was my source for the public domain photos included herein. This collection can be better viewed at two sites: Экспедиция полковника В.М.Вонлярлярского на Чукотский полуостров в 1900 году – Российская Национальная Библиотека – Vivaldi (nlr.ru) and Ėkspedit͡sii͡a polkovnika V.M.Vonli͡arli͡arskogo na Chukotskiĭ poluostrov v 1900 godu. | Library of Congress (loc.gov).

TO THE FAR NORTH

Introduction

"I'm sitting in my cabin admiring a guard's gleaming bayonet—we've been taken prisoner. Strictly speaking, this ain't bad for the start of the 20th century," young Ivan Nikolaevich Akif'ëv sardonically wrote in his diary for 21 August (n.s.) 1900, making light of an irretrievable breakdown in relations that had just occurred between Americans and Russians aboard a ramshackle ship in the North Pacific, a breakdown that had nearly led to deadly violence. Despite his black humor, Akif'ëv was unwittingly prescient, for violence narrowly averted would indeed be "not bad" during the decades that followed 1900. This adventurous physician's diary of his remarkable journey from St. Petersburg across Europe, the Atlantic, and the United States and then around the North Pacific offers a unique glimpse into the dawning of the tumultuous twentieth century—a glimpse on the one hand peripheral to Russia but on the other hand very much a reflection of it. Nonetheless, *To the Far North* is unlike any other source on the topic because its topic is unique.

We know little about Akif'ëv (pronounced a-keef-YOFF). Although his diary appeared in two editions,[1] it was apparently not popular. This has left its author largely forgotten. The only other sources on him are a pair of obituaries

---

1. See "Notes on the Translation."

and an anonymously authored website.[2] From these we learn Akif'ëv was born 29 August 1872 in the city of Nizhnii Novgorod. Following grade school he enrolled in the medical program at Moscow University, where his interests also included politics, industry, and philosophy. While still in university, Akif'ëv was named to lead a medical unit assigned to the government's construction of the Samara–Zlatoust railway. After receiving his doctorate, Akif'ëv visited Italy and Switzerland, where he met the Swiss physician and medical researcher Emil Theodor Kocher, who in 1909 would win a Nobel Prize for his pioneering work in aseptic surgery. In 1898, despite youth and inexperience, though possibly through connections he made on the railway project, Akif'ëv joined a timber-surveying expedition to Korea organized by Vladimir M. Vonliarliarskii and Aleksandr M. Bezobrazov.[3] Akif'ëv's connection to Vonliarliarskii proved auspicious. After their return to Russia, the latter chose Akif'ëv as his family's physician, and the 1900 Chukotka Expedition that Akif'ëv joined and is documented here was realized thanks to Vonliarliarskii's efforts.

This expedition's precedents included the Alaska Gold Rush, which itself formed part of a global phenomenon of Industrial Age gold rushes. Moreover, in 1897 Russia had adopted the gold standard and, although this temporarily devalued the ruble, "the state may have improved the connections between the Russian and Western capital markets, allowing . . . private borrowers to obtain funds more plentifully, more cheaply, or both," writes one historian.[4] But just as importantly, the 1900 expedition reflected Russia's anxiety over American encroachments in Chukotka, not just to find gold but to trade with

2. V. V. Korsakov, "Bezvremenno pogibshaia sila," *Russkie vedomosti*, 23 April 1906, 4; "Skorbnyi listok," *Voenno-meditsinskii zhurnal* 216 (June 1906): 396. See also nlr_spb, "Bezvremenno pogibshchaia sila," *LiveJournal*, 19 May 2016 "Безвременно погибшая сила": nlr_spb—LiveJournal, accessed 31 August 2022. The website is vague as to its sources.

3. This expedition amounted to what one historian calls a "front" that allowed St. Petersburg to send military servicemen in the guise of timber surveyors to the Yalu River valley, where during September–October 1898 they reconnoitered march routes for a future war against Japan. David S. Crist, "Russia's Far Eastern Policy in the Making," *Journal of Modern History* 14, no. 3 (1942): 318–21; Karen A. Snow, "Commercial Shipping and Singapore, 1905–1916," *Journal of Southeast Asian Studies* 29, no. 1 (1998): 48; Olga Crisp, "The Russo-Chinese Bank: An Episode in Franco-Russian Relations," *Slavonic and East European Review* 52, no. 127 (1974): 212; Vladimir Tikhonov, "Korea in the Russian and Soviet Imagination, 1850s–1945: Between Orientalism and Revolutionary Solidarity," *Journal of Korean Studies (1979–)* 21, no. 2 (2016): 394; and Thomas C. Owen, "Chukchi Gold: American Enterprise and Russian Xenophobia in the Northeastern Siberia Company," *Pacific History Review* 77, no. 1 (2008): 49, 55–56. Though appointed the expedition's physician, Akif'ëv surveyed and even authored for a secret government publication a description of one potential march route. For his service he was awarded the Order of St. Stanislav (or Stanislaus), third class (Korsakov, "Bezvremenno pogibshaia sila"; "Skorbnyi listok"; and "'Bezvremenno pogibshchaia sila'").

4. Ian M. Drummond, "The Russian Gold Standard, 1897–1914," *Journal of Economic History* 36, no. 3 (1976): 688. See also Olga Crisp, "Russian Financial Policy and the Gold Standard at the End of the Nineteenth Century," *The Economic History Review* 6, no. 2 (1953): 158, 168–69.

the native Chukchi. The Alaska Gold Rush drew many thousands to the Far North to look for riches. Many prospectors did get rich, yet a greater number arrived too late to stake a claim, and some even ended up broke and stranded in Alaska.[5] American newspapers and government reports alerted St. Petersburg to the fact that some miners were turning their attentions across the narrow Bering Strait to Chukotka, where Russians had almost no physical presence.[6] Anglo intervention into North Pacific regions claimed by Russia had been going on for decades. The Chukotka Expedition was therefore designed as much to reassert Russian sovereignty over the peninsula as it was to find gold.[7]

Enter Karl Ivanovich Bogdanovich (Karol Bohdanowicz) (1864–1947), a russified Pole, ethnographer, and geologist working in the Russian government's Mining Department (*Gornyi departament*). In spring 1898, while researching on Kamchatka Peninsula, he theorized on the basis of Alaska's and Chukotka's similar geology that gold might also be found in Chukotka. In late 1899 Bogdanovich seems to have joined a group that we know pressured the Ministry of Agriculture and State Domains (MZGI[8]) to recommend an expedition be launched to ascertain the presence of both gold and foreigners on Chukotka.[9] But the government under Nicholas II was an indecisive one and lacked the wherewithal to act on the MZGI's recommendation.

---

5. Pamela Cravez, *The Biggest Damned Hat: Tales from Alaska's Territorial Lawyers and Judges* (Fairbanks: University of Alaska Press, 2017), ch. 1; Bathsheba Demuth, "Geology, Labor, and the Nome Gold Rush," in *A Global History of Gold Rushes*, ed. Benjamin Mountford and Stephen Tuffnell (Berkeley: University of California Press, 2018), 252–72; Bathsheba Demuth, *Floating Coast: An Environmental History of the Bering Strait* (New York: W. W. Norton, 2019), 202–15; T. H. Carlson, "The Discovery of Gold at Nome, Alaska," *Pacific Historical Review* 15, no. 3 (1946): 259–78; Andrea R. C. Helms and Mary Childers Mangusso, "The Nome Gold Conspiracy," *Pacific Northwest Quarterly* 73, no. 1 (1982): 10–19; and Leland H. Carlson, "Nome: From Mining Camp to Civilized Community," *Pacific Northwest Quarterly* 38, no. 3 (1947): 233–42.

6. Robert McGhee, *The Last Imaginary Place: A Human History of the Arctic World* (Chicago: University of Chicago Press, 2007), 146–49; John J. Stephan, *The Russian Far East: A History* (Stanford, CA: Stanford University Press, 1994), 33–39, 86–88; Adam Burns, *American Imperialism: The Territorial Expansion of the United States, 1783–2013* (Edinburgh: Edinburgh University Press, 2017), 42–4; and *Zabytaia okraina: rezul'taty dvukh ekspeditsii na Chukotskii poluostrov, snariazhennykh v 1900–1901 gg. V. M. Vonliarliarskim, v sviazi s proektom vodvoreniia zlotopromylennosti na etoi okraine* (St. Petersburg: Tipografii S. Suvorina, 1902), 15–16.

7. Demuth, *Floating Coast*, 63, 86–87; Nikolay Ivanovich Kulik and Anastasiya Alekseevna Yarzutkina, "Gold of Chukotka and Foreign Investments: Institutional Approach," *Middle-East Journal of Scientific Research* 15, no. 3 (2013): 409.

8. MZGI stands for Ministerstvo zemledeliia i gosudarstvennikh imushchestv.

9. K. I. Bogdanovich, *Ocherki Chukotskago poluostrova* (St. Petersburg: Tipografiia A. S. Suvorin, 1901), vii–iii; Sergei Iu. Vitte, *Vospominaniia*, 2 vols. (Moscow: Izdatel'svto sotsial'no-ekonomicheskoi literatury, 1960) 2:279; and V. M. Vonliarliarskii, *Chukotskii poluostrov. Ekspeditsii V. M. Vonliarliarskogo i otkrytie novogo zolotonoskogo rainona, bliz ust'ia p. Anadyria, 1900–1912 gg.* (St. Petersburg: Tipo-litografii K. I. Lingarda, 1913), 12–3. We learn from Akif'ëv's diary that Bogdanovich returned to Russia from Kamchatka aboard the same vessel that brought Akif'ëv back from the 1898 Korea Expedition. Therefore, they knew each other before the 1900 expedition.

At this point, Vladimir Mikhailovich Vonliarliarskii (1852–1946) decided to step in. The son of a major general and a member of the Novgorod elite who earned renown during the Russo-Turkish War (1877–78), Vonliarliarskii was an entrepreneurial investor with some experience in mining. Together with Bezo-brazov he had launched the 1898 Korea Expedition in the expectation that war with Japan was imminent. The historian W. E. Mosse cites the so-called Bezo-brazov Clique as an example of Russia's "reckless adventurism in the Far East."[10] Now, in 1900, having somehow learned of the government's internal debate over Chukotka, Vonliarliarskii offered once more to fund a venture that he intended would benefit both himself and Russia. On 15 March the government accepted Vonliarliarskii's offer, on the condition that Bogdanovich be named the expedition's leader. Bogdanovich and Vonliarliarskii immediately began planning for an excursion to the other side of the world, one to begin in spring 1901, but fast-developing events would move this start date ahead by a full year.[11]

Time and money were of the essence. American prospectors seemed ready to inundate Chukotka the same way they had Alaska. With the clock ticking, the Russians needed to get there before "tens of thousands of Americans crossed the Bering Strait upon the first opportunity, to establish mines on our side."[12] Vonliarliarskii, devised a complicated investment scheme with British and American partners, to finance the expedition. Principal among them was the Englishman Frederick W. Baker. The Russian government approved their arrangement but further stipulated that Bogdanovich, already named to lead the expedition, should also serve as an MZGI representative with plenipotentiary authority to address any predatory mining by foreigners. The granting of such power to a state geologist spotlights the near absence of communications between the metropole and its Far North periphery, but also shows that for St. Petersburg, the expedition's primary goal was to assert territorial sovereignty. Further indicating a suspicion of foreigners, the final government contract specified that *Russian* employees were to be hired in numbers sufficient to offset the expedition's foreign hires, and that the expedition include a physician, three Pacific Fleet sailors, and a detachment of five "Kamchatka Cossacks" (i.e., mixed-race Kamchadal irregulars).[13] Akif'ëv was Vonliarliarskii's

10. W. E. Moss, "Imperial Favourite: V. P. Meshchersky and the Grazhdanin," *Slavonic and East European Review* 59, no. 4 (1981): 544.

11. Vonliarliarskii, *Chukotskii poluostrov*, 13; Vitte, *Vospominaniia*, 2:279; and Bogdanovich, *Ocherki*, viii–ix.

12. Bogdanovich, *Ocherki*, ix.

13. Bogdanovich, *Ocherki*, x; *Zabytaia okraina*, 18–19; Kulik and Yarzutkina, "Gold of Chukotka," 410; Vonliarliarskii, *Chukotskii poluostrov*, 13–14, 16, 20; and Jane Woodworth Bruner, "The Czar's Concessionaires: The East Siberian Syndicate of London; A History of Russian Treachery and Brutality," *Overland Monthly* 44 (1904): 412.

obvious choice as physician. But the addition of armed Cossacks and sailors gave the expedition a military aspect that had negative consequences, as readers of this diary will learn.

Vonliarliarskii and his partners finalized plans hastily. Upon learning no ships were available in Vladivostok, they agreed that Baker's American contacts would acquire and outfit a steamer in San Francisco instead. Traveling by rail and sea, the Russian expeditionaries (along with Baker and the American George Roberts, who were to join them in London) would rendezvous with the foreign expeditionaries in California, where common laborers and a steamer crew would also be hired. On reaching the Chukotka Peninsula, the expedition would begin its investigations and then rendezvous with the *Iakut*, a Russian navy ship that would deliver additional supplies as well as the government-mandated Cossacks and sailors. The *Iakut* would then depart for its yearly trip to Kamchatka before returning to pick up the Russian nationals and bringing them to Vladivostok. Everyone else would return to San Francisco. These logistics, though they proved imperfect, are nonetheless admirable for their time: they coordinated participants, travel routes and means, and supplies over an expanse that stretched from St. Petersburg west to the Bering Sea. Not wanting to lose another minute, the Russians left St. Petersburg on 18 April / 1 May, one day after Vonliarliarskii signed his contract with the Mining Department.[14]

Akif'ëv begins his diary with an account of their train crossing the frontier. He was twenty-seven years old and keen for adventure. By his side was Aleksandr Gennad'evich Miagkov (1870–1960), from Kazan, a young nobleman who as a university student at the St. Petersburg Technology Institute had joined public demonstrations, run afoul of the authorities, and left before getting his degree.[15] His relationship to Akif'ëv before the Chukotka Expedition is unknown, but clearly they were friends from the beginning of the trip. In 1901 Miagkov would publish his own much briefer memoir of the expedition, one that aligns closely with Akif'ëv's diary.[16] Akif'ëv mentions Miagkov frequently, often using the affectionate diminutive "Gennad'ich." Their friendship, combined with the contents of the diary and what we know of Miagkov's background, suggests that despite his evident service to the tsarist government, Akif'ëv's personal politics skewed left. His previous forays to Europe and Asia undoubtedly exposed him to other views. Whereas many educated Russians during this period espoused liberal beliefs, his diary shows

14. *Zabytaia okraina*, 19; Bogdanovich, *Ocherki*, x; and Vonliarliarskii, *Chukotskii poluostrov*, 1–2, 20.

15. [Anon.], "Miagkov Aleksandr Gennad'evich (1870–1960)," in *Religioznaia deiatel'nost' Russkogo Zarubezh'ia*, Aleksandr Gurevich, dir., Tsentr religioznoi literatury VGBIL, Biblioteki-fonda Мягков Александр Геннадьевич (narod.ru), accessed 4 May 2022.

16. A. Miagkov, "V poiskakh za zolotom," *Russkoe bogatstvo* 8 (1901): 102–59.

Akif'ëv to have been especially broadminded. Nonetheless, his diary (at least) does not suggest he was an intellectual. Rather, it reveals interests typical of most Euro-American males his age: Akif'ëv expresses an interest in women and mentions more than once drinking with Miagkov and other buddies; he goes hunting and sport shooting but also enjoys gathering wildflowers.

More historically significant, Akif'ëv's diary demonstrates a worldview shaped by nationalist, imperialist, and racialist discourses. He stereotypes (in order) Germans, Brits, Jews, Americans, Chukchi, and Japanese. His praise of Russians is at times chauvinistic, though he never rises to the level of the evangelizing Slavophiles. What he thought about Americans before the expedition can only be guessed. But based on what he says about those who accompanied him to Chukotka, he seems to have shared the prejudice held by the Moscow merchant Pavel A. Buryshkin, who condemned Western businessmen's motives as decidedly more mercenary than those of their Russian counterparts. Buryshkin embodied a general Russian contempt toward America as being a country debased by capitalist excess, which antisemites and polonophobes furthermore blamed on immigrant Jews and Poles.[17] Akif'ëv was therefore perhaps conditioned to see the worst in the American expeditionaries, though if his and Bogdanovich's accounts are to be believed, those Americans did their best not to disappoint. Still, it is difficult to know if Akif'ëv is representative of other Russians who shared his socioeconomic position at the time, given that the scholarly literature focuses on Slavophiles and other exceptional groups.[18]

Like now, worldwide anxiety characterized the period when Akif'ëv penned his diary; and like now, nationalism, imperialism, and racialism factored into that anxiety. Nationalism emerged in Europe during the late eighteenth century. Following the 1789 French Revolution, Giuseppe Mazzini and others on the political left embraced and promoted nationalism as both a domestic and an international pacifier. But as the century wore on, nationalist politics increasingly slanted right. One's personal identity came to be associated with one's ethnicity and language; those most obsessed with these signifiers increasingly aligned themselves with conservative, reactionary, and repressive agendas.[19] This latter form of nationalism, the kind most recognizable to us today, was not limited to Europe: versions of it emerged in the Americas and Asia as

---

17. Thomas C. Owen, *Russian Corporate Capitalism from Peter the Great to Perestroika* (New York: Oxford University Press, 1995), 121–22, 127.

18. Robert V. Allen, *Russia Looks at America: The View to 1917* (Washington: Library of Congress, 1988); and Hans Rogger, "America in the Russian Mind: Or Russian Discoveries of America," *Pacific Historical Review* 47, no. 1 (1978): 27–51.

19. For both a classic and a more recent treatment, see E. J. Hobsbawm, *Nations and Nationalism since 1780: Programme, Myth, Reality* (Cambridge: Cambridge University Press, 1992); and John Connelly, *From Peoples into Nations: A History of Eastern Europe* (Princeton, NJ: Princeton University Press, 2020).

well. A foremost example is Japan, where, after the Meiji emperor resumed supreme power in 1868, his government promoted an aggressive nationalistic chauvinism to unify the country against predatory Great Powers.[20]

Imperialism's precise origins and contours remain contested (some historians even demote imperialism to "modern colonialism"), but it seems fair to say that imperial practices—whether enacted by Japan, Great Britain, or the United States—carried associations with nationalism and international commerce, and sometimes also with religious or secular evangelism.[21] For example, some in America's newly emerging middle class saw themselves as "the true police of the world," dutybound to protect civilization and prevent other countries' citizenry from turning into "savages like the communal tribes of the Aleutian Islands," whereas certain Russian Slavophiles had "a faith that Russian principles—whether communal, religious, or socialist—pointed the way to the future of mankind."[22]

Like warm air to a typhoon, popular racism and the racialistic pseudosciences of social Darwinism, Lombrosianism, and degeneration theory blended with nationalism and imperialism to form "the popular belief that each nation embodied a particular race."[23] Enjoying widespread embrace by northern-tier peoples, racialism found purchase even among defenders of such multinational empires as Russia and Great Britain and led to the proverbial conceit of "The White Man's Burden."[24] "Inferior races," Lord Acton therefore wrote, "are raised by living in political union with races intellectually superior. Exhausted and decaying nations are revived by the contact of a younger vitality."[25] For his part, Nikolai Kh. Bunge, writing in 1895 as chairman of Russia's Committee of Ministers, shared this belief: "The weakening of the racial differences in the borderlands can only be achieved by attracting the core of the Russian population to the borderlands, but this will only work if

---

20. Nancy K. Stalker, *Japan: History and Culture from Classical to Cool* (Berkeley: University of California Press, 2018), 209–22.

21. Norman Etherington, "Reconsidering Theories of Imperialism," *History and Theory* 21, no. 1 (1982): 1–36; Robin A. Butlin, *Geographies of Empire: European Empires and Colonies c. 1880–1960* (New York: Cambridge University Press, 2009), 21–30; Raymond F. Betts, *The False Dawn: European Imperialism in the Nineteenth Century* (Minneapolis: University of Minnesota Press, 1975), 122, 147; and Jane Burbank and Frederick Cooper, *Empires in World History: Power and the Politics of Difference* (Princeton, NJ: Princeton University Press, 2010), 287.

22. James Weir Jr., quoted in Michael McGerr, *A Fierce Discontent: The Rise and Fall of the Progressive Movement in America, 1870–1920* (New York: Free Press, 2003), loc. 1416, Kindle; Rogger, "America in the Russian Mind," 31.

23. Lloyd S. Kramer, *Nationalism in Europe and America: Politics, Cultures, and Identities since 1775* (Chapel Hill: University of North Carolina Press, 2011), 150–51.

24. The White Man's Burden—The Kipling Society, accessed 20 June 2023.

25. [John Dalberg-Acton], *Home and Foreign Review* 1 (July 1862): 17.

the Russian population does not adopt the language and habits of the border-lands but rather brings their own there."[26] Racialism instrumentalized geopoliti-cal rivalry as a struggle for survival and fueled a second colonial thrust, aimed this time at Africa as well as Asia. By virtue of their subjugation, colonized peoples were deemed racially inferior and subjected to an "epistemology of ex-ploration": ethnography and anthropology early established markers that distin-guished between the "barbarian" and the "civilized" worlds.[27]

Imperial Russia played a role in all these developments. By 1900, chauvin-ism, xenophobia, antisemitism, and nativism were evident throughout the Russian government and society.[28] Like everyone at the time, Akif'ëv was sub-jected to the discourses conveying these beliefs. Despite this, Akif'ëv tries in his diary's pages to give voice to his more humanitarian instincts. Sometimes he succeeds, sometimes he does not. Though manifestly proud to be Russian, Akif'ëv's patriotism is tempered, as his withering criticisms of his government and society show, especially when he complains about the failure to develop the Far North and condemns the Sakhalin penal colony. He also offers measured praise for both Japanese and Chukchi culture. Akif'ëv's intuitive (one might add, *medically refined*) appreciation of other peoples' humanity therefore ap-pears alongside the reigning discourses he grew up with and digested. His di-ary embodies Akif'ëv's own humanity as well as his era's anxieties, and as such, it documents a tension that moderns still confront today.

This book's afterword will say more about the Chukotka Expedition and its participants during the period following that covered by Akif'ëv's diary. For now, let us turn to the diary itself, which consists of eleven chapters that can be grouped into five sections: (1) Akif'ëv's journey as far as San Francisco and his time there; (2) the Chukotka Expedition and its denouement (this forms the bulk of the diary); (3) Akif'ëv's and Miagkov's visit to the Sakhalin penal

26. Quoted in Anatolyi Remnev, "Siberia and the Russian Far East in the Imperial Geography of Power," in *Russian Empire: Space, People, Power, 1700–1930*, ed. Jane Burbank and Mark von Hagen (Bloomington: Indiana University Press, 2007), 442.

27. Butlin, *Geographies of Empire*, 121–23, 225–26.

28. Andreas Renner, "Defining a Russian Nation: Mikhail Katkov and the 'Invention' of National Politics," *Slavonic and East European Review* 81, no. 4 (2003): 659–82; Walter Laqueur, *Black Hundred: The Rise of the Extreme Right in Russia* (New York: Harper Collins, 1993); Don C. Rawson, *Russian Rightists and the Revolution of 1905* (New York: Cambridge University Press, 1995); and Robert Wein-berg, *Blood Libel in Late Imperial Russia: The Ritual Murder Trial of Mendel Beilis* (Bloomington: Indiana University Press, 2014). Jeffrey Brooks argues that by the early twentieth century, Russians were be-coming *less* xenophobic. He cites voting patterns for Duma candidates as evidence of this. I believe that much of the evidence, including that cited here and Akif'ëv's diary itself, counters this interpreta-tion. *Pace* Jeffrey Brooks, "Official Xenophobia and Popular Cosmopolitanism in Early Soviet Russia," *American Historical Review* 97 (1992): 1432–33.

colony; (4) their brief stay in Vladivostok and sea voyage to Japan; and (5) their visit to Nagasaki and its environs.

The first section offers valuable insights on fin-de-siècle life in Europe and especially in the United States. Akif'ëv visited what were then the United States' three largest cities: New York, Washington, DC, and San Francisco.[29] What he saw intrigued him. Like most foreign travelers, Akif'ëv uses his home country as his standard of reference and evaluates the differences as either good or bad, as tourists will. This section also offers a valuable account of steamer and train travel, showing the state of these technologies in 1900.

The second section concerns the Far North—a flexible term in both English and Russian (*dalekii sever*, or more recently, *dal'nii sever*), though one that generally refers to the Arctic and sub-Arctic regions. More narrowly (given the distinction sometimes made between the "Near North" and the "Far North"), it designates everything north of the tree line running across Eurasia and North America. The Far North is often portrayed as barren, bleak, cold, and inimical to human existence, and Akif'ëv more or less follows this trend.[30] In Nome, Alaska, during January, the daily mean temperature is 5.2°F, the average high 13.1°F, and the average low –2.8°F. On the Chukotka Peninsula in, say, the village of Uélen, it is colder still. There February, not January, is the coldest month: the daily mean is –5.3°F, the average high 0.9°F, and the average low –11.2°F. Uélen remains cool during its brief summers: July's daily mean is 44.1°F, and its average high only 50.2°F.[31]

Little other than scrub grass and lichen grows on the peninsula named after the indigenous Chukchi, a people linguistically and genetically similar to the Inuit. Until the late nineteenth century, when substantive trading with Americans and Europeans varied their diet, the Chukchi existed almost exclusively on protein derived from aquatic life or domesticated reindeer. Those along the coast (the *nuunamiut*) lived a more communal and sedentary lifestyle than those inland (the *tareumiut*). Both groups practiced shamanism, developed advanced hunting techniques, and maintained relations, though taboo prevented them

---

29. Their respective populations at the time were 3,437,202; 278,718; and 342,782. US Census Bureau, 1900–1980, *Population of the 100 Largest Cities and Other Urban Places in the United States: 1790 to 1990*, Working paper no. 27 (2003) No. HS-7. Population of the Largest 75 Cities: 1900 to 2000, https://www.census.gov/history/pdf/los_angeles_pop.pdf, accessed 20 April 2022.

30. On cultural constructions of the Far North, see Louis Rey et al., eds., *Unveiling the Arctic* (Calgary: Arctic Institute of America, 1984); Julia Augusta Schwartz and Vilhjalmur Stefansson, *Northward Ho! An Account of the Far North and Its People* (New York: Macmillan, 1929); and Veli-Pekka Tynkkynen et al., eds., *Russia's Far North: The Contested Energy Frontier* (New York: Routledge, 2018), chaps. 13–15.

31. "Pogoda i klimat" Климат Уэлена - Погода и климат (pogodaiklimat.ru), accessed 1 September 2022; National Weather Service Zone Area Forecast for Southern Seward Peninsula Coast (weather.gov), accessed 1 September 2022.

intermarrying. Chukchi were just one among dozens of peoples native to Beringia who traded, intermarried, learned from, and warred against their neighbors. After Russians and Cossacks ventured into northeastern Siberia, they committed numerous atrocities attempting to subdue the Chukchi, who stoutly resisted and sometimes massacred Russian encampments in turn. By the late eighteenth century an informal truce allowed both sides to trade without excessive bloodletting. Russians valued the furs and walrus tusks the Chukchi could supply, whereas Chukchi relished the tea, tobacco, and hardware the Russians provided. But interaction with Europeans and Americans forever altered and nearly destroyed Chukchi society. It was the Russians who apparently brought syphilis and influenza to the Chukchi, whose reindeer herds were similarly stricken by a disabling hoof disease probably introduced by horses. The anthropologist Vladimir G. Bogoras visited Chukotka around the same time Akif'ëv did, and he learned that smallpox had been ravaging the Chukchi for decades and that an outbreak in 1884 reduced their numbers by a third.[32]

These circumstances should be kept in mind when reading Akif'ëv's comments on the Chukchi. Despite the Academy of Sciences having appointed him and Miagkov the expedition's ethnographers, neither appears to have been trained in the discipline and certainly they had no previous firsthand knowledge of the *nuunamiut* they encountered. Rather, they represent those privileged white male dilettantes who flitted around the globe under imperialism, documenting various "savages" and their strange cultures and bringing home cultural artifacts for personal collections and museums.[33] Like Lord Acton

---

32. Waldemar [Vladimir] Bogoras, "The Chukchi of Northeastern Asia," *American Anthropologist* 3, no. 1 (1901): 80–108; James Forsyth, *A History of the Peoples of Siberia: Russia's North Asian Colony, 1581–1990* (Cambridge: Cambridge University Press, 1994), 18, 53, 71–74, 79–81, 145–50; Yuri Slezkine, *Arctic Mirrors: Russia and the Small Peoples of the North* (Ithaca, NY: Cornell University Press, 1994) 17, 28, 62; Stephan, *Russian Far East*, 87; Robert Sasso and Merwyn Garbarino, *Native American Heritage* (Long Grove, IL: Waveland Press, 1994), 104; John H. Relethford, *Reflections of Our Past: How Human History Is Revealed in Our Genes* (Boulder, CO: Westview Press, 2003), 143; Emilio F. Moran, *Human Adaptability: An Introduction to Ecological Anthropology* (Boulder, CO: Westview Press, 2007), 32, 126–27; John P. Ziker, *Peoples of the Tundra: Northern Siberians in the Post-Communist Transition* (Long Grove, IL: Waveland Press, 2002), 65; Marina Mogilner, "Russian Physical Anthropology of the Nineteenth–Early Twentieth Centuries: Imperial Race, Colonial Other, Degenerate Types, and the Russian Racial Body," in *Empire Speaks Out: Languages of Rationalization and Self-Description in the Russian Empire*, ed. Ilya Gerasimov et al. (Boston: Brill, 2009), 155–89; and Demuth, *Floating Coast*, 52–56, 64, 85–98, 150–51. Accounts from Chukchi perspectives are rare. For a novelized but highly informative account written by a modern-day Chukcha about life during the time Akif'ëv visited, see Yuri Rytkheu, *A Dream in Polar Fog*, trans. Ilona Yazhin Chavasse (New York: Archipelago Books, 2005).

33. In addition to the sources already cited, see Hellen Tilley and Robert J. Gordon, eds., *Ordering Africa: Anthropology, European Imperialism and the Politics of Knowledge* (Manchester: Manchester University Press, 2017); Richard Lee and Karen Brodkin Sacks, "Imperialism and Resistance: The Work of Kathleen Gough," *Anthropologica* 35, no. 2 (1993): 181–93; David Arnold, ed., *Imperial Medicine and Indigenous Societies* (Manchester: Manchester University Press, 2017); Gustavo Lins Ribeiro, "World

writing about Britain's colonial subjects, Akif'ëv characterizes the Chukchi as racially inferior and requiring Russian tutelage. He essentializes them, apparently without being aware that their society had recently collapsed.[34] By the same token, Akif'ëv expresses admiration for Chukchi culture and describes outsiders' negative influence on them. His treatment of several sick Chukchi also demonstrates his fundamental humanitarian disposition toward them.

Similarly contributing to our historical and anthropological understanding of the Far North is Akif'ëv's inimitable description of Nome at the height of its gold rush. He vividly portrays conditions in this boomtown, ones that typify this period's gold rushes, and he dilates in particular on the addiction his expedition's Americans had to the precious metal. Akif'ëv's character sketches are affecting, enabling him to deploy his steamer's dipsomaniacal Norwegian American captain Edward Jahnsen for comic relief. By focusing on the expedition's messy and scandalous dissolution, this section also demonstrates the superficial presence of law and order in the Far North. Our diarist goes on to detail his visit to Petropavlovsk, a once important fortified outpost now reduced to a shadow of its former self. Unlike his descriptions of other places, here Akif'ëv shows considerable knowledge of Petropavlovsk's past and offers a bit of a history lesson, though he gets some details wrong. Chekhovian vignettes of local Cossacks and the expedition's hard-drinking laborers enliven this section.

Akif'ëv and Miagkov next ventured to Sakhalin, a description of which forms the diary's third section. The penal colony was infamous both inside and outside Russia as a devil's island comprising thousands of murderers guarded by sadistic officials. Scandal sheets exaggerated the horrors there, but Sakhalin was nevertheless a telling example of tsarist penology's shortcomings. Anton Chekhov's 1895 book and similar accounts served in part to attract adventurers who wanted to do a bit of slumming on the island.[35] The

---

Anthropologies: Anthropological Cosmopolitanisms and Cosmopolitics," *Annual Review of Anthropology* 43 (2014): 483–98; and Bruce Kapferer, "From the Outside In," *Social Analysis: The International Journal of Anthropology* 62, no. 1 (2018): 27–30.

34. Some scholars have claimed that Russian anthropologists' attitudes were less racialized than those of other countries, but too much contrary evidence seems to exist, including that found in Akif'ëv's diary. For a discussion on this debate, see Marina Mogilner, "Beyond, against, and with Ethnography: Physical Anthropology as a Science of Russian Modernity," in *An Empire of Others: Creating Ethnographic Knowledge in Imperial Russia and the USSR*, ed. Roland Cvetkovski and Alexis Hofmeister (New York: Central European University Press, 2014), 81–120.

35. Anton Chekhov, *Sakhalin Island*, trans. Brian Reeve (London: Alma Classics, 2019). See also Andrew A. Gentes, trans. and ed., *Eight Years on Sakhalin: A Political Prisoner's Memoir* (New York: Anthem Press, 2022); Andrew A. Gentes, trans. and ed., *Russia's Penal Colony in the Far East: A Translation of Vlas Doroshevich's "Sakhalin"* (New York: Anthem Press, 2009); and Andrew A. Gentes, *Russia's Sakhalin Penal Colony, 1849–1917: Imperialism and Exile* (New York: Routledge, 2021), ch. 11, 12, 22.

nine days Akif'ëv spent visiting its two main posts were sufficient for him to bear witness to this particular "imperial ruin," to borrow Ann Laura Stoler's words.[36] In addition to providing information about the penal colony, this section offers the best example of Akif'ëv's willingness to criticize Russia and its institutions: he unreservedly denounces the penal colony's maleficence.

Before they could visit Japan, Akif'ëv and Miagkov had to go to Vladivostok to catch a different steamer. Founded in 1860, home to Russia's Pacific Fleet, and with a population of just under thirty thousand, Vladivostok was by far the largest Russian city on the Pacific.[37] Like San Francisco, it consists of several steep hills rising from the edge of a beautiful natural harbor. At the time, Vladivostok had perhaps the most diverse population of any city in Russia. Besides various nationalities from all parts of the empire, large numbers of Chinese, Korean, and other foreign immigrants were present. Asians accounted for the bulk of Vladivostok's population and the city had a slew of foreign consulates. "A dozen languages echoed in local stores, banks, hotel lobbies, and brothels," writes John Stephan.[38] Akif'ëv shares that he knew the city from at least one previous visit, which he would have completed in tandem with the 1898 Korea Expedition. This brief interstitial fourth section offers further details on the fraught nature of both sea travel and hotel accommodations on the periphery.

The diary's final section chronicles Akif'ëv's and Miagkov's weeklong stay in Nagasaki, Japan. As with Vladivostok, Akif'ëv had visited Nagasaki before. It shared close ties with Vladivostok. Seventy percent of the Russian Far East's immigrant Japanese population originated from Nagasaki, and many owned businesses in Vladivostok. An undersea telegraph cable connected both cities, and Nagasaki was a primary destination for Russian tourists. Between 1878 and 1894, it served as a coal depot and winter port for Russia's Pacific Fleet. At the time of Akif'ëv's visit, Nagasaki's Russian population of 177 was the largest of any Japanese city. Its hotels and resorts were international destinations, and not just for Russians.[39] Despite this foreign influx and Japan's rapid

---

36. Ann Laura Stoler, "Considerations on Imperial Comparisons," in Gerasimov et al., *Empire Speaks Out*, 50–55.

37. 1897 census, Institut demografii imenii A. G. Vishnevskogo, Higher School of Economics, Moscow Демоскоп Weekly - Приложение. Справочник статистических показателей. (demoscope.ru), accessed 18 April 2022.

38. Stephan, *Russian Far East*, 71–80, 84, 86.

39. Stephan, *Russian Far East*, 57, 76, 84; and P. Vaskevich, *Ocherk byta Iapontsev v Priamurskom krae* (Verkhneudinsk: Tipografiia A. D. Reifovicha, 1905), 5, and table, p. 5; also published as P. Vaskevich, "Ocherk byta Iapontsev v Priamurskom krae," *Izvestiia Vostochnago Instituta* 15, no. 1 (1906): 1–30. See also Yukiko Koshiro, *Imperial Eclipse: Japan's Strategic Thinking about Continental Asia before August 1945* (Ithaca, NY: Cornell University Press, 2013), 61–62, and tables, pp. 51, 53.

Westernization, life in Nagasaki remained wedded to tradition, as Akif′ëv's descriptions of its inhabitants and the royal heir's ceremonial arrival there show. He brings the city and its lush natural surroundings to life in these, his most poetic passages; and it is in this final chapter that he best conveys his love of travel and appreciation of foreign cultures. Nonetheless, Akif′ëv concludes his diary on a rather sour note, having wandered from a description of ill behavior by some Japanese in Korea to making racialized generalizations about the Japanese people as a whole.

If viewed objectively, Akif′ëv's diary can be appreciated as a historical document that emphasizes how moderns cognitively navigate life using discursive structures created by both our predecessors and our contemporaries. Simply put, we are all, like Akif′ëv, shaped by the culture(s) in which we are reared and reside, and so we apprehend the world in sometimes radically different ways from each other. Appreciating that we are as much victims as progenitors of history, we might hesitate before passing judgment on our predecessors. Akif′ëv's *To the Far North* reveals one young man's growing awareness of the world and cultures different from his own. Taking as a given that such awareness expands upon later reflection, we can assume that Akif′ëv's experiences led to a greater awareness of himself. But how much he truly gathered from his global travels can only remain a matter of speculation, for Akif′ëv's life was to be tragically cut short.

Chapter 1

From Petersburg to New York

18 April/1 May 1900. At last, I'm fulfilling my desire to go to the Far North, and the train is now flying over the rails like a bullet carrying us from Petersburg. We're aiming west to go to the far northeast. We're heading for London so as to sail from there across the Atlantic Ocean, cut across North America, and then, from San Francisco, to leave for the Chukotka Peninsula. That is, in order to go from one end of Russia to the other, we've chosen to go through Europe and America and to sail across two oceans.

"Is this really the most convenient?" we were asked.

Without a doubt. Actually, were we to take a route through the Indian Ocean on a Volunteer Fleet steamer,[1] it would take around 50 days to get from Petersburg to Vladivostok. Whereas going from Petersburg to San Francisco via America takes all of around 20 days. Vladivostok and San Francisco are in practically the same position for going to upper Chukotka. Hence, were we to go through Vladivostok we'd lose nearly an entire month. I should say, i.e., as of now, there's still no land route connecting Vladivostok, so it's more convenient to go *from* Russia *to* Russia *through* America. Two years ago, I had to get from Petersburg to Vladivostok as quickly as possible. I was actually trav-

---

1. Literally, "na nashem dobrovol'tse," referring to the Russian Volunteer Fleet (Russkaia dobrovol'naia flota) that was established during the 1870s to ship goods, convicts, and the occasional freeman from Odessa to the Russian Far East.

eling for nearly a month, and it was costly. Quite expensive, in money *and* in health.

So, we're going to the Chukotka Peninsula to find gold. Two years ago, a bounty of gold, wonderful gold, was discovered on the Alaska coast near the Nome cape. Extracting straight from the shore was possible with no tools other than a spade and a trough; of course, this generated gold fever and people quickly filled up the coastline near Cape Nome. Since Alaska is separated from the tip of Chukotka by merely the width of the Bering Strait, and, moreover, gold has been found on Kamchatka and the shores of the Okhotsk Sea, it was proposed gold might be found on the Chukotka Peninsula.

Some Americans have tried to find gold there. The results of their searches aren't properly known, but the subject of gold on the Chukotka Peninsula reached the London and New York exchanges and interested capitalists. Foreigners showed up in Petersburg and appealed to the Russian government for a legal concession on the Chukotka Peninsula.

Mr. V–ii [Vladimir Mikhailovich Vonliarliarskii] appealed similarly, and he, as a Russian, was given preference over the foreigners. He's obtained a concession to mine and extract gold on the whole of the peninsula for a term of 5 years; if he establishes a business and his works are productive, the concessionary term will be extended. V–ii decided to accept foreign capital into his company, and this is why he's doing well. The Chukotka Peninsula shores are completely undefended, and our warships visit once every few years, so it's very easy for foreigners to access them and, clearly, predatory parties of American gold prospectors might penetrate there, and small Russian expeditions might collide with the predators and given their weak resources, be unable to dislodge them. Since foreign capitalists have affiliated themselves with the company it's been possible to hope they'll consider it in their own interests to protect their concessionary area from predatory invasion.

The Englishman [Frederick W.] Baker, who's concluded a three-year agreement with V–ii, is one such affiliate. Together, they decided on and devised an expedition. The expedition would be mixed: the expedition leader had to be a Russian engineer, and Russian assistants and workers were to accompany him, save for an English engineer and his two assistants.[2] Baker took on the

---

2. The foreign engineer and assistants who joined the expedition were Roberts, Rickard, and Landfield (see Dramatis Personae). Of these three, only Rickard, who was anyway Swiss, had an association with England. At this point, Akif'ëv seems not to have known the nationalities of all those who were to join the expedition. But he makes similar mistakes regarding other foreign expeditionaries, even after spending weeks with them, which suggests he had difficulty parsing anglophone accents. These mistakes will be noted as they arise.

responsibility of personally hiring a vessel for the expedition. It was decided to charter a ship in San Francisco and take the ship from there to the North.

That's the other reason we're traveling through America.

Six of us are on the way to London, in addition to our concessionaire V–ii and his wife, who will travel from there to the Paris Exposition.[3]

I was also counting on going to the exposition, but instead I'm now going to the North.

My traveling companions are as follows. The expedition leader is the mining engineer Karl Ivanovich Bogdanovich, famed for his travels through Central Asia and Kamchatka. A year ago, he returned from an expedition to Kamchatka, where he'd spent three years aboard a French vessel on which I was also returning from Korea. We became acquainted there. Karl Ivanovich is blond and short, with an intelligent, very nervous face. Evidently, his expedition did not pass without affecting his health. He is quiet, but occasionally brightens and speaks very interestingly. With him travels his wife, his inseparable companion during his travels. It is remarkable that she, only having just returned from three years in the snows of Kamchatka and enduring many difficulties and deprivations that undermined her constitution, because of which she cannot be called fully healthy, is once more traveling north, knowing of the bad weather, cold, and all sorts of dire deficiencies and inconveniences that await her there. Yet, for the moment, she is very excited and cheerfully preparing herself for the long trip.

Traveling with me alongside them is the former mining student Aleksandr Gennad'evich Miagkov, a tall, healthy young man with considerable experience prospecting for the railroads in Russia and Siberia and now enormously lucky to be going on the expedition. Making the same impression is our new acquaintance, Al'fred Markovich Makferson, traveling with us in the capacity of interpreter and jack-of-all-trades. He is a tall, sinewy, gray man, albeit wholly energetic and winsome. He's an Englishman by nationality and a trained mechanic, born in Petersburg, educated in England, well-traveled, and fluently and correctly speaking five languages in addition to his own. He's hearty, forceful, cheerful, and loves company, and we were soon chatting with each other over a few glasses of wine.[4]

---

3. The Paris Exposition lasted from mid-April to mid-November 1900. All the same, Vonliarliarskii changed plans and (presumably with his wife) ended up accompanying the group as far as New York City. As Akif'ev mentions below, besides himself, also arriving in New York were his friend A. G. Miagkov, their lackey Nikolai, A. M. Makferson, and K. I. Bogdanovich and his wife. (See Dramatis Personae.)

4. The English spelling of Makferson's name was probably either McPherson or MacPherson. However, Akif'ev refers to him throughout by his name and patronymic and treats him as a fellow Russian. For this reason I transliterate Al'fred Markovich's surname as "Makferson" and have included him among the Dramatis Personae's "Russians."

Also traveling with us is a certain Terek Cossack,[5] the Ossetian Nikolai S., in the capacity of trusted fellow. He was also in the earlier expedition in northern Korea, where he proved his reliability. Finally, there's me—the expedition's doctor and a great travel enthusiast.

We're in a big hurry so as not to waste an hour's time; for this reason, our train is the *Nord Express*, i.e., the priciest train.[6] A ticket from Petersburg to London costs 150 rubles, though it beats the usual courier train by 12 hrs. and in terms of transfers. The car is very well appointed, though the twin compartment is very cramped. The train includes a restaurant car, though prices are quite high. Speaking for myself, it would generally have been better to leave one day earlier aboard the courier.

21 April/4 May. We've been here in London almost a full twenty-four hours now. I'm sitting in the *reading room* of the *Charing Cross Hotel* and recalling these past two days.[7]

On the morning of the 19th, we crossed the border into Eydtkuhnen, and there drank our first mug of beer in a foreign land.[8] We transferred to German cars, which proved far more comfortable than the Russian. Aleksandr Gennad'ich and I occupied an entire four-person compartment, bright, with large windows.

Just as we were setting off, the waiter came and brought us breakfast. From the first dish it was evident we were no longer in Russia. We got small sardines and chunks of sausage to eat and, later on, sausage again, chunks of pike, and even pieces of some sort of half-cured meat, for around 5 Marks. Expensive, but not exorbitant. A half-bottle of beer cost 80 Pf. (40 kop), and this in Germany, land of beer and sausage. In a word, more expensive and far worse than in our country. The same waiter tried to shortchange Al'fred Markovich; he was quickly found out but then tried to shortchange us, but we proved him guilty. He was left extremely dissatisfied and went away with a look of insulted innocence. He'd been trying to cheat us with unusual aplomb and an insouciant look. So much for German scruples and honesty.

We traveled through Germany for a whole day and reached Belgium. At the border there, our baggage went entirely uninspected, and we were simply asked: "Where are you going?"

---

5. The Terek Cossack host was named after the Terek River in the Caucasus.

6. Established by a French company in 1896, the *Nord Express* ran from St. Petersburg to the French Channel ports, where many travelers transferred to steamers to continue onward to England and America.

7. Akif'ëv writes both "reading room" and "Charing Crosse [*sic*] Hotel" in Latin letters.

8. Today Eydtkuhnen is called Chernyshevskoe and located within the detached Russian province of Kaliningrad.

"To London."

"Very well"—and that concluded the inspection.

I was more impressed by Belgium than Germany. Perhaps because it wasn't the first time I'd visited the latter, or maybe it was the truly boring, uniform plain and the standardized plowed parcels evenly divided by highways. The trees along those roads are spaced evenly. Nowhere will you see there a smattering of beautiful individually designed huts. A solitary hut can be seen no place there. In general, Germany usually reminds one of an enormous chessboard. Given these images, if you close your eyes and imagine drawing a picture of the German landscape and then open them an hour later, the picture you've drawn will almost certainly correspond to the actuality. Everything is extremely standardized. The Germans themselves are nauseatingly undifferentiated. Whomsoever you see, each has the same shaven face and mustachios unnaturally curled and puffed up to resemble cat whiskers, the same rigid shoulders, circular barreled chest, unusually insolent and measured gait, and that Olympian something in his person as is in our private bailiffs and police chiefs. A completely uniform mass. In a word, a train through Germany may be just as boring as a train through, for example, our Baraba Steppe.[9]

Belgium makes a rather similar impression, but the landscape is more varied there, mountains are encountered in places, and the train passes through tunnels, though the population is even denser: cities are here and there. Small gardens with a dozen gaunt, stunted trees in front of the houses make for a curious sight. They resemble the little gardens our children make when playing: a small space is cleared, twigs piled in rows. That's what it's like there. It's so strange that I wouldn't want to live there, regardless of the culture. It's all good, even very good, but probably too good for an unsophisticated Russian fellow. I'd be bored there, like Sadko with the Sea Tsar.[10] You'd stand there, limited by that restrained culture and chartered environment, remembering our endless steppes, the thick forests of Siberia and the Urals, and the magnificent peaks of the Caucasus and the Altai.

In Brussels, we had to change seats because the train was going straight to Ostend whereas our tickets were for Calais. Since there was absolutely no other car, we relocated to the dining car. It was terribly crowded and very uncomfortable. Entering the car, I found on the floor near one of the armchairs a monastic rosary in the form of the Holy Mother of Lourdes. I appealed to those present, but no one had lost it. "No," they all answered in the negative, it was my fortune and would bring me luck in my travels.

---

9. The Baraba Steppe is a vast plain in southwestern Siberia.

10. In Russian mythology, a musician named Sadko visits the court of the Ruler of the Seas.

Around 4 o'clock the train reached the coast at Calais, where the small ferry steamer onto which we transferred was already rocking about. An hour later another train arrived from Paris with a load of passengers. The steamer ended up completely full of people: the sons and daughters of foggy Albion were filling up the hold, cabins, and deck. These islanders, especially the women, were remarkably plain and unattractive. The young women's bodies were tall and thin as sticks, without any shape, and their movements were also as sharp and angular as their bodies. Their faces were angular, like those of birds or horses, with extended sharp noses, long teeth, and red hair; their expressions were bitter. The men were better looking and of solid height and healthy, but their reserved manner was repulsive: something like a mannequin. Onto the steamer our baggage went, similarly uncontrolled without inspection and entirely without customs stickers. We had barely left shore when the steamer began rocking and waves began washing over the deck. I moved further toward the prow and, finding a covered spot, sat and watched the green waves angrily assault the deck. The rocking increased and one lady after another began blanching and disappearing below, where something resembling a cholera ward had already formed. Groans and still more inharmonious noises could be heard from all sides. After several minutes the men joined this choir, and it was strange to see how tortured these renowned seafarers and island dwellers were, whereas the residents of the mainland and steppe land, completely unused to the sea, remained healthy and calm.

An hour and a half later we were already in Dover. "Farewell, mainland," and tomorrow we'll be saying: "Farewell, Old World!"

Final impressions of the Old World have been good: I like England. From Belgium, where every piece of land is used or plowed so there was nowhere we could stretch out, we've suddenly ended up in a land of husbandry. Meadows segmented by rows of trees stretched alongside the railway. The meadows were even and covered in lush grass. Sheep, horses, and cows grazed hither and yon in them, and all this livestock was thoroughbred and splendid.

Large areas have also been devoted to the cultivation of hops, which for the most part go to producing porter and ale here. The population is not as dense as in Belgium, though the train rushed through small cities and settlements every now and then. The structures are very unique and quite beautiful. Apartment buildings are not tall, typically two or three floors, are quite small, and their walls above the first level are separated into two or three sections. Each section represents a separate apartment. In this there's a difference between the Continent and England. In Europe, our apartments usually occupy a single floor, or, more accurately, are located within a single level, and were half an apartment to be on one floor and half on another, it wouldn't be considered very convenient. In England, it's the reverse: apartments of several rooms occupy all the levels. Chimneys are situated in a building's partitions, resulting in an entire series of chimney pipes in the roof, as if it's covered with bristles.

We reached London late in the evening; we rode through the city for a long time but could see nothing in the darkness. We're staying, as I've already said, in the *Charing Cross Hotel* located near the train station of the same name. Our great amount of baggage was also admitted without inspection. Two are sharing one room with me. Al'fred Markovich has his own bed, but Aleksandr Gennad'evich and I have to share a single huge bed. Yesterday morning, Aleksandr Gennad'ich and I, with our *Baedeker* in hand, began sightseeing, since we had just one day to relax and see the city. The city is gigantic, and to see it, you'd have to spend some time in it; therefore, I got no impression other than that the city is large and foggy.

23 April/6 May. Here we are in the Atlantic Ocean. We crossed Europe like lightning, and yesterday the train bolted us from London to Liverpool, where we boarded a steamer.

Our ship is the Cunard Company's huge four-deck steamer *Compania*. It's 610 ft. long and has a maximum speed of 22 ½ knots.[11] It's like a gigantic apartment building. The steamer makes an absolutely dizzying impression. It is superbly appointed: located within the spacious dining area below deck are five long dining rooms, along the sides of which are tables for 10 people each. We're occupying one such table. The entire dining hall is made from a beautiful wood and covered in carpets. Our dining room includes a music hall. At the center of the dining room is a lantern around which the furniture has been arranged. All the furniture, as well as the piano and organ, are made of some sort of white wood, like Karelian birch, and this also decorates the dining room walls. Located on an upper deck are a smoking and a reading room. The latter is situated in the prow, so sitting there during rough seas is not especially pleasant. The deck for first-class passengers is large and expansive. It costs 4 shillings for the privilege of using an armchair during the crossing. Everything would be good; yet, with regard to the cabin, that most urgent of issues, nothing can be said of comfort. We were a trio: myself, Aleksandr Gennad'ich, and Al'fred Markovich had been assigned such a small cabin the three of us could barely stand up at once. There were beds for two, but the third had to sleep on the divan beneath the porthole. And for this cabin they're getting

---

11. Launched in 1892, the RMS *Compania* was, like her sister ship the *Lucania*, preeminent among the Cunard Line at that time. It was actually 622 feet long, displaced 18,450 tons, and could carry 2,000 passengers. It sank in 1918 after colliding with a British warship.

75 pounds sterling from us, i.e., more than 700 rubles for a passage of 6 days. The cost is astonishingly high, and it would seem passengers might get better comfort for such a price. We ended up appealing to Baker, who's traveling with us and seems to have pull here. Thanks to his help, Al'fred Markovich was assigned a separate cabin, and we shall remain a duo.

Our company has grown. Having joined us is V–ii's colleague Baker, an elegant, polite, middle-aged man, but clearly prepossessed with himself. Then there's the American, [George D.] Roberts, a crippled 72-year-old man with a wedge of gray beard and eyes nearly blind but, they say, a sage of the gold business. He's traveling in the capacity of consultant. He's said to have three times made up to 6 million dollars and three times to have lost it. Now he has nothing and evidently wants this expedition to correct the matter. Next to him sits a young man, quite pleasing by appearance, the American [John W.] Dowlen[12]; he's traveling in the capacity of assistant engineer to [Forbes] Rickard,[13] whom we're to meet in San Francisco: what he's like is not known. Finally, there is Mr. [E. L.] Stern, a very pleasant elderly gentleman with noticeably softly feline mannerisms, exceptionally obliging but, judging from his eyes and the expression on his face, a Jew. What his relationship is to the expedition is unknown.[14] We're all sitting at one of the separate tables, saying little to each other since we don't know English and they don't know Russian or French, save for Baker, who's speaking French.

29 April/12 May. Here we are, about to reach America. Well, what can be said? The ocean did not pamper us: there was a proper storm on the third day and last night as well, so that instead of 500 miles we made only 350 to 400 per day. We will just make New York tomorrow evening, thank God.

---

12. Akif'ёv transliterates his name as "Даулен"; Bogdanovich as "Доулин". Jane Woodworth Bruner misidentifies Dowlen as both "Dolan" and as being English. Jane Woodworth Bruner, "The Czar's Concessionaires: The East Siberian Syndicate of London; A History of Russian treachery and Brutality," *Overland Monthly* 44 (1904): 412.

13. In her article, Bruner misidentifies him as "Forbes Pickard, of Denver," and further confuses him with fellow expeditionary "Mr. Pickard, of London" (see Dramatis Personae). (Bruner, "Czar's Concessionaires," 412.) But in describing preparations for the expedition, Bogdanovich mentions "the min[ing] eng[ineer] Forbes Rickard, an Englishman, my former first assistant." (K. I. Bogdanovich, *Ocherki Chukotskago poluostrova* [St. Petersburg: Tipografiia A. S. Suvorin, 1901], 1–2.) However, the full extent of the two men's relationship before the Chukotka Expedition is not known. Another source identifies Rickard as being born in Switzerland in 1867, emigrating to England, graduating from London's Royal School of Mines, and moving sometime during the 1890s to Denver to work in the Colorado mining industry. He also worked in British Columbia and Mexico. The specific dates for his time in Russia are not known. "Biographical Note," Forbes Rickard Papers, WH333, Western History Collection, The Denver Public Library, https://archives.denverlibrary.org/repositories/3/resources/8645, accessed 13 July 2021.

14. Stern would serve as the expedition's purchasing agent in San Francisco.

On the morning of the third day, I went up top and was literally amazed by the ocean's severe, awesome look. Gigantic gray waves rushed furiously toward our steamer, hoping to crush and break it, but it adroitly rose to their crests and, from there, dove deep, deep. They were saying this morning that it was descending such that the waves were covering everything and slowing progress to a crawl. The wind was tearing the summits off the waves, rivulets of cold saltwater were pouring onto the deck and through the windows. The steamer was groaning, creaking, and shedding water everywhere; not just those in the topmost of the steamer but everyone, from the top down, kept getting inundated. Up top, as if in mourning, the wind howled piercingly through the tackle. Every passenger could hear the disturbing sounds through their cabins. Aleksandr Gennad'ich was so sick he couldn't go up top. The most unpleasant thing was sitting or lying in one's cabin and feeling the ship drop rapidly from the crest of a wave. It felt like the floor was disappearing beneath your legs, like your bunk was falling and you were falling somewhere with it.[15]

Now and then, something strange and very unpleasant was taking place in your head and stomach. Per my own experience, I know it's better when seasick either to walk or, if already lying down, to lie motionlessly and try to fall asleep. The many who are unable to eat or drink while seasick suffer most of all. Aleksandr Gennad'ich was similarly unable to eat or drink and tried, but was unable, to eat an orange. I'm now laughing that he's been so pampered and spoilt he couldn't even tolerate oranges. You have to make yourself eat, and if the food doesn't stay down—you eat again. I always do this and feel fine.

During those five days I even managed to fatten up. Speaking personally, fattening up wasn't difficult with the following regimen. At 9:30 there was a signal inviting you, as you wished, to—*breakfast*. You ate however much and whatever you wanted, the choice of dishes being enormous, though it couldn't be said which was going to be very tasty. (In my opinion, no kitchen in the northern hemisphere can compare with our kitchens—e.g., on the Volga steamers or at Testov's in Moscow.[16]) So, you thoroughly gorged yourself in the morning. At 12 o'clock they brought you chicken or beef bouillon. An hour later, you gorged again on a limitless supply of dishes—this was *luncheon*. At five o'clock there was tea, and at seven there was supper. As such, the whole day was spent eating. There haven't been many diversions. Evenings here have usually proceeded just as mechanically and drearily as the days. The ladies, having wrapped their legs in their plaids, lie in armchairs on deck and the men

---

15. The voyage was indeed rough. A passenger reportedly said: "It was a winter's voyage in May." This became a story headline. "A Winter's Voyage in May," *New York Tribune*, 13 May 1900, 7.

16. Ivan Iakovlevich Testov's restaurant in central Moscow was renowned for its traditional Russian cuisine.

smoke and drink whiskey in the *smoking-room*. The drink *Apollinaris whiskey* is remarkably appealing among them: they drink it before and after a meal, in a word, at anytime of day or night, and some drink it more or less constantly.[17]

Yesterday, a charity concert was held for a collection of 60 pounds sterling—that is, i.e., 600 r[ubles]. Were we to hold a charity concert for such a collection, and were it not filled from beginning to end with all sorts of celebrities, they'd say: "No luck." Nowadays, Russians are very demanding, incline toward criticism, and really can't get around without it. Of course, this has its good sides: we have stricter expectations regarding singing and music, and among us these are far superior. Yet here, they were listening to their own not very talented male and female singers with genuine pleasure and applauding them with enthusiasm! I was asked to sing a Russian song as well and had agreed, but my accompanist Aleksandr Gennad'ich was at that point collapsing and falling asleep, so I didn't participate.

We found entertainment in the smoking lounge. They were betting on the steamer's speed. It must be said there was an entire company of Jews going around, among whom there was even one elephantine townsman who fled military conscription and has been living in America for 25 years now. This company arranged the game. It consisted of this: the steamer covers an average of 480 miles per day. They placed 20 slips above and below this figure. Everything above the number of the last slip was sold separately, as well as what was below. But because the quantity of slips was limited but those wishing to bet were many, the slips were auctioned off and the price increased 5 or more times. At noon the following day, the distance known to have been covered during the previous 24 hours would be ascertained and whoever had the ticket with that figure on it got the money. There were still other kinds of games. Ten slips, numbered zero through nine, were sold. The winner was whoever had the same numeral as the last one in the mileage covered the previous day.

I observed three games, and the Jews won all three times. Really, you'd have said the Jews here are quite the experienced seafarers, or so it seemed.

There's been nothing of greater interest, and there's no going on deck today: the wind is terrible and cold; spray lashing the deck. The usual decorous conversations are in the salon. Whiskey and cards are in the smoking lounge. The one good thing is that we land soon, and I'll see the New World for the first time.

---

17. This was a simple rickey, nothing more than bourbon with Apollinaris mineral water on the rocks.

CHAPTER 2

In America

New York—Washington [DC]—San Francisco—Outfitting the expedition

5/18 May. I'm tired; I've undressed but want to record my impressions before going to sleep. They're so numerous they've gotten mixed up, and I don't have the time to write them down. We've been here in Washington for three days already, but tomorrow we're moving on. Of New York I'm left merely with an impression of noise and commotion. We got there at night. Thousands of lights, which would stop in place and then move, covered the entire river. Steamers of rather unique construction were crossing from one shore to the other by the minute. Above the decks were some sort of beams that moved. They were connected to pistons by vertical cylinders. People, brightly lit seemingly by lanterns, filled the steamers.

We were lingering offshore. Our giant was unable to freely turnabout, so it hovered near the wharf; two small steamboats rested themselves against it, one on the prow, the other on the stern, and helped guide it; in such way, the *Compania* was turned around. The enormous customs building was full of people. There was a great amount of baggage, and it couldn't be released because they'd dreamed up a very thorough inspection: you took a letter and marked all your baggage with this letter. The porters loading the baggage brought it to a hall where the letters were posted on a wall in alphabetical order and put your baggage beneath the corresponding letter. Of course, it so happened that amid the bustle they made mistakes, but nevertheless there was in general an orderliness to the passengers and their baggage.

We were finally released at midnight and went to the hotel. Our hotel—the Waldorf Astoria—is the best in New York. It was very nice: clean, organized, and there were baths, elevators, music halls, huge dining rooms, and corridors through which many of the guests were always strolling. Everything was wonderful, but rather uncomfortable. It quickly felt constricting. In the dining room where, practically without exception, all the men were smoking cigars and cigarettes, our female traveling companion also began smoking a cigarette after dinner. A steward immediately came over and told Al'fred Markovich, as our leader, that this was not accepted there. A strange affair. Our ladies, though not many, freely smoke where they like, but here ladies can't smoke in public, even though very many smoke at home, and they also drink in secret. The price for a room and victuals was quite significant, but the dining couldn't be said to be very tasty. The next day, Aleksandr Gennad'ich, Al'fred Markovich, and I found a small eatery nearby and dined and supped there, with pleasure.

The next morning we allowed ourselves to go to the dining room and met Dowlen there. He showed us a newspaper that had come early that morning. Our arrival, our expedition, and, moreover, much that was incorrect and superfluous were written about there. Having seen this newspaper, Karl Ivanovich became angry and spoke about it to Baker, who gave a dissatisfied look. "If Dowlen is saying this" (he suspected him for some reason), "then I'll remove him from his position without discussion." But I'm sooner ready to suspect this article was written based on Roberts's words.[1]

We occupied ourselves with roaming the city. Strictly speaking, nothing distinguished it from large commercial cities. Only, there were huge multistory apartment buildings sticking out like teeth among the others there. We counted 26 stories in one. It wasn't attractive, and I suppose living on the 26th floor wouldn't be especially pleasant, though height means nothing there: every apartment building has an elevator. The streets were filthy, and it could be

---

1. Upon the group's arrival in New York, Roberts (probably at Vonliarliarskii's urging) gave an impromptu press conference. "[M]any Englishmen and Americans are interested" in the concession granted Vonliarliarskii, he told the *New York Times*, adding: "Some expert mining engineers have explored some of the territory covered by the concession, and their reports assure us that the supply of gold there is *inexhaustible*." Roberts said Bogdanovich would lead the expedition, and he boasted, "[W]e shall carry with us the most perfect machinery for the class of mining necessary." This machinery plus the low cost of Russian labor would allow them to mine gold in Chukotka cheaper than in Alaska, he asserted. ("To Seek Gold in Siberia," *New York Times*, 14 May 1900, 1 [emphasis added].) Subsequent developments suggest Roberts's comments were intended to raise the cost of shares in the East Siberian Syndicate, a company Vonliarliarskii and Baker had earlier formed to finance the expedition. The issue of market speculation is further discussed in this book's afterword.

imagined they're never swept, and the traffic was terrible. Cabs flew one after the other down the street. The racket and din of the wheels, the ceaseless bell-ringing, the cabbies' flashing eyes, all this deafens and stuns the unhabituated person. Railcars travel not just over bridges along several series of rails but even overhead, on rails laid across columns. Enormous, pachyderm-shaped horses pull gigantic, terrifying carts, and people cover the pavements. You're simply astonished by how, in the perspectival distance where several lines converge, all this doesn't get confused or crushed together and by how a policeman can, with a single wave of his magic wand, halt in a moment all the movement and obligingly guide pedestrians across the street and even escort a lady by the arm.

Aleksandr Gennad'ich and I considered leaving there to see Niagara Falls, but the circumstances were so complicated that Karl Ivanovich invited us all to go to Washington instead. It was a great pity to be denied this excursion, but there's no bad without the good and I'm satisfied to have now reached Washington. This is the best city of all I've seen so far.

Washington has been built geometrically, so looking at a schematic is enough to easily orient yourself. The city center—the Capitol—is a hill on which wonderful buildings have been constructed—the *Parlament*[2] and the library. Extending straight from it are streets to the north, south, east, and west—the so-called Capitol Streets, etc. Then there are the streets stretching North to South and designated by a series of numbers, beginning on either side of the Capitol, and those stretching W—E and lettered in alphabetical order. In addition, in a radius around the Capitol run boulevards—the avenues.

We're staying at the Raleigh Hotel on Pennsylvania Avenue. Ours is the liveliest of streets. But for this it's charming. Though there's lots of traffic, there's neither a crowd nor a crush here. Electric trains, nearly always full of people, go back and forth here and there. Practically no cabbies are visible, and velocipedists hasten one after the other. Apparently, there's no other city with such a lot of velocipedists. And this is understandable: the streets are wide, covered with asphalt, and flat as a table, and we can't hear any fuss or racket. And the velocipedists here ride exceptionally well. I watched one little girl, apparently 8 or 9 years old, riding a velocipede without holding the handlebar and tacking with unusual dexterity between pedestrians and tramcars. They sometimes even carry small loads on velocipedes. In addition, there's a mass of electrical carriages here.[3] Especially nice is that trees gird all the streets here, sometimes in several rows. Even on commercial streets there's the scent of

---

2. I.e., the Capitol Building. Akif'ëv writes this in Latin letters as a misspelled reference to the French *Parlement*.

3. Electric automobiles.

greenery and flowers. Were the city to be viewed from the height of a bird in flight it would appear suffused in green.

With this goal we dragged ourselves to the top of the obelisk honoring Washington and admired the view from there. The Potomac River was visible to the south, a beautiful ribbon that girdles the city and is joined by another flowing from the west and hidden in the green distance. The obelisk stands in a park opposite the famous White House. The park is a delight. It covers the entire expanse from the obelisk to the Capitol. Beyond it there appeared a small garden with green lawns, hedges, tree-lined alleys, and a row of beautiful red buildings. These buildings are the museums. We inspected them in detail, as much as we could.

The Smithsonian Institute has two floors. Archaeology occupies the upper. The many contributions there on the history of America are very interesting. The lower floor is a zoological collection: the fish, birds, and shellfish are rather impoverished. The national museum has several sections—the geological is abundant and visually well arranged. The zoological is very poor, and there are nautical, ethnographic, and a section for all sorts of souvenirs. In the medical museum there's an abundant collection of preparations for pathological anatomy and collections of skeletons and of plaster casts of skin diseases. Even though everything has been gathered in a single place, and though much there is good, much of it isn't very interesting. It makes for a very muddled impression.

A certain Jew we've become acquainted with brought us to the Corcoran picture gallery as well.[4] He asked if we'd seen anything similar in our Russian picture galleries. In fact, in ours, we've never encountered such impoverishment. Several statues standing there proved to be plaster copies of the European originals. The European paintings were far from interesting. There was almost nothing but what was completely mediocre. We did find one interesting piece there. This was a picture of Aivazovskii's depicting the joy with which the Russian folk greeted the delivery of American grain during the famine year. In the foreground was a troika and on it, a chap with an American flag was rushing down a street filled with people. After the Aivazovskii, we felt rather awkward.[5]

---

4. Possibly E. L. Stern (see Dramatis Personae).

5. Ivan Konstantinovich Aivazovskii (1817–1900) was a Romantic painter best known for his seascapes. The painting Akif'ev refers to is called *Razdacha prodovol'stviia* (Distribution of Provisions) and commemorates the arrival of American aid during the 1891–92 famine in Russia. This painting so offended the Imperial Court's patriotic sensibilities that it forbade its showing in Russia, so Aivazovskii brought it to the United States and donated it to the Corcoran Gallery. Jacqueline Kennedy later arranged for it to be loaned to the White House. In 1979 the painting was sold to a private collector. In 2008 it was auctioned off to a sponsor who then returned it to the Corcoran. As of this writing, the

We also visited the national library and the parliament on the Capitol. We went to the parliament building first. We were let in with tickets that our Jewish acquaintance kindly got for us. To us, it was strange to see how casually the parliamentarians conducted themselves. One just lay on a couch smoking and reading a newspaper until another began speaking from the rostrum. It was like he was simply at home. The library is a marvel of architecture. But much of it is empty hallways. In fact, the showcases contain some terrible nonsense. Generally, if European and American museums are to be compared, ours have a lot of things but are not nice buildings, whereas here, there is practically nothing inside of luxurious, marvelous buildings.

Now a few words about the theater. Aleksandr Gennad'ich and I, since we understand little English, went not to the dramatic theater but to the vaudeville. We did this once in New York and twice in Washington. The cynicism was quite unabashed and there was no gracefulness whatsoever. Strictly speaking, these theaters more closely resemble farces: not in terms of construction—no, they're well-built and comfortably appointed—but the performers and the program really are a farce. There's folksinging, ventriloquists, gymnasts, very talented jugglers, humorous couplets, chansons, the cancan, and vaudeville. In a word, everything you'd want. But what of the singing, what of the dancing? Strictly speaking, the singing can truly be called shameful, and the dancing is a swinging of legs, in truth very fast but in no way graceful, and rather deformed.

At almost all the vaudevilles someone jumps out a window because there's such a racket being made, as if ten shelves of glassware are being smashed or crockery cupboards tipped over. Summoned to this terrible mess, young and old rush to form the audience, and they shout and clap like lunatics. Yet they're even more pleased when there's a skit involving a boy or a girl of 7 who starts knocking out the cancan and singing couplets of dubious content. During the conclusion, the vision of a cancanning family of mulattoes whips the audience into savage delight. The husband cancans astonishingly indecently with his wife and, in exact imitation of father and mother, their children, a boy and a girl, do the cancan. A wonderful upbringing!

They sit with their hats on in the theaters, and smoke in the hallways and even drink there. And here's something especially unique—they whistle as a sign of approval here, and sometimes they whistle terribly piercingly. In general, I got a poor impression of the tastes of the American public.

---

painting remains banned in Russia, indicating that the awkwardness Akif'ëv and Gennad'ich felt after seeing it was a matter of their wounded national pride.

We leave Washington tomorrow and I'm very sad to be parting from it. It's so nice in the evening under the electric lights, when along a bright smooth street as if covered in snow the velocipedes' and electric machines' multicolored lights fleetingly appear. There are many people out, and no wonder: the evenings here are splendid and warm, the air filled with the smells of trees and flowers. In a word, it's very nice.

There is one bad thing—and that is the hotel staff. They serve so sluggishly and unwillingly and while doing so give a look as if they're doing you a great favor. It's said they don't take tea in America. Absolutely not true. It's brought and gladly accepted all the time, even more than among us in Russia!

9/22 May. The train has halted this morning on the plain without having reached Rawlins Station.[6] Aleksandr Gennad'ich woke me at six o'clock this morning and I, quickly dressing, left the coach and paused, blinded by the beauty of the view as it revealed itself. Directly in front of me stretched a bare steppe, a field only sparsely covered by bushes and tufts of skinny grass, and beyond it the gigantic white teeth of dazzling, snowy mountain peaks rising into the blue sky. Having been stopped inside the stuffy car for several minutes, it was nice to breathe in the clean, fresh air.

It turned out that four versts ahead of us a tender had jumped the rails and was blocking our way. One of our steam engines and another one that appeared from somewhere were assisting it, allowing us to roam the steppe. Little by little, one after another, passengers began alighting from the train and dawdling about the steppe. From the first look at the track it was clear the rails hadn't been well laid. Most important, nowhere did they seem to be even. The sleepers' ends were hanging in midair and touching the ground only in the middle. No thought was given to stone ballasts here. Indeed, what can be said of a stone ballast when here and there the sleepers have rotted nearly all the way through? And you should have seen the fasteners! Many of the plates were broken or hadn't been solidly hammered down in places, and the spike-head had a very small radius and, in spots, was turned sideways so it wasn't meeting the rail. This was where the tender's rear wagon jumped the rails and went along the sleepers, revealing that most of the sleepers were rotted.

The steam engines couldn't get the tender onto the tracks, and they turned back. After a little while, a maintenance train with a crane arrived from the

---

6. Lit., "do stantsii Raulin"—apparently Rawlins Station, Wyoming. The Union Pacific station there did not become operational until 1901, however, so Akif'ëv may be referring to a provisional station.

station. We ran to watch. It was quite interesting when the whole corner of the tender was lifted into the air and then placed correctly onto the rails. Within a few minutes everything was satisfactorily done.

Today we breakfasted in some station for 75 cents and were fully satiated. Now we're sitting, calculating the train's speed. Our train proves to be going around 70 versts per hour.[7] The trip is well organized and also follows such a beautiful railway. It's getting terribly expensive. We're traveling in style, in Pullman express cars. During the day it's a normal coach with divans, but at night they convert it for sleeping and each person gets a long, wide bunk surrounded by drapes. This allows men as well as ladies to sleep inside it. The train has dining and smoking rooms. The food's not bad, though pricey.

The railway from Washington to Chicago passes through wooded areas. After Chicago, it changes character. There stretches a rolling plain reminiscent of the steppe in Samara Province, though further on there's a desert, a real sand desert sparsely covered with short, prickly bushes, and you don't see any signs of life there, just emptiness and lifelessness. Only, there are distant mountains' snow-covered peaks rising on the horizon. The air in the cars is awful, hot, and thoroughly suffused with dust particles unnoticeably percolating up from all the chinks in the car. We're traveling through those areas where the railway was built under military guard, but should you want to see the Indians that were here then, they're absolutely invisible.

We've gone only 8,000 ft. into Echo Valley in the state of Utah, but we're breathing more easily.[8] It's very beautiful here. The valley is bordered by tall, precipitous, red-colored cliffs, and here and there enormous turrets and rock walls can be seen. There's a whole series of pillars and separate rocks resting, by some miracle, on their small bases. A gust of wind through the valley would seem to knock them over. Running through the valley is a stream, over which we've crossed several times.

10/23 May. Desert again, even more lifeless than yesterday. There's absolutely no water. It's hot, there's dust in our eyes, grit in our teeth, and our bodies itch terribly. We're bored and foul tempered. Today, Al'fred Markovich read to us an article from a San Francisco newspaper about the Russians, gold in Siberia, and how the Russians are trying to attract American capital. Named were Grand Prince Nikolai and Prince Dolgorukov, whose ancestors supposedly founded Moscow, and it was reported that V–ii's son was traveling, but

---

7. Around 46 miles per hour.
8. Akif'ëv appears to be referring to Echo Canyon, in what is today called Echo State Park just east of Salt Lake City.

not a single one of our names was there. In a word, a typical American newspaper.[9]

26 May/8 June. We've now been living in San Francisco for nearly two weeks. We're staying in San Francisco's Palace Hotel, again in one of the best areas, and preparing ourselves for departure. These past days there's been much that's new, quite unforeseen, and not entirely welcome.

As soon as we arrived, Stern kindly took us to see the steamer. Our disappointment was total. Speaking personally, I could not believe it when Stern told us: "And here's our steamer!" It seemed so small and unappealing to us. It was only 150 feet long and the only passenger cabins were on the stern deck, making for a rather strange sight. There was a deck in the middle of the steamer, two cabins had been built on the prow for the sailors, and there were two levels of passenger cabins on the stern. The steamer's middle looked like it was broken. A strange dirt was everywhere. All this didn't exactly please us, but Stern was lauding it: "What a delight for a steamer!" I returned to the hotel and told Karl Ivanovich: "I don't like that steamer!" He went to see it and came back. "Not true," he said, "it's a good steamer." "Very well," I thought, "later, we'll see."

Karl Ivanovich met here our old sailor friend Lemashevskii, who's sailed the Pacific Ocean for thirty years, and invited him to view the steamer. He looked at it and said: "The steamer's good and solid."

Well, if this old sea wolf says so, then of course you have to trust him.

We consented to agree and sign the contract, moreover, Stern earnestly promised it was impossible to find a steamer here but that he'd luckily found the *Samoa*. "You know, all the steamers are going to Alaska, to Nome," he said. "They're all full, so we need to grab the *Samoa*."

The circumstances were hopeless and had to be agreed to, though the price was enormous. It is five thousand dollars a month, and the coal as well as maintaining the ship's crew will be charged to the expedition. In a word, given as much as it's costing the expedition to rent it, it would have been better still to purchase a steamer.

Baker, who was to conclude the contract, arrived several days later. The contract was signed in Karl Ivanovich's name, but not as V–ii's agent and the expedition leader but as a proxy for the "East Siberian Syndicate."[10] Aleksandr

---

9. Vonliarliarskii had three sons, none of whom accompanied the expedition. Concerning American capital, see "To Seek Gold in Siberia." Nikolai Nikolaevich Romanov (1856–1929) was a grandson of Nicholas I. At the time Akif'ev was writing, at least three princes Dolgorukov were alive. Neither the Romanovs nor the Dolgorukovs are known to have had a financial stake in the 1900 Chukotka Expedition.

10. The East Siberian Syndicate was the investment company formed by Vonliarliarskii and Baker to finance the expedition; its shares were sold on the London and New York exchanges.

FIGURE 2.1.   The *Samoa*

Gennad'evich and I weren't privy to this business and so didn't pay attention to it, though it seemed rather strange to us. We've been busy either with purchasing everything needed for our journey or with getting to know San Francisco.

In this, our new acquaintance has been very helpful. Dowlen, one of our imminent travel companions, has introduced us to his young wife. We've found her very pleasant, and her behavior towards her husband has been so sweet and endearing that we've taken to him with great affection and found them friendly company. Nearly every evening, after having dined and spent a very cheerful time with them in a restaurant, we attend the theater. Especially amusing is that the spouses Dowlen speak only English and we are from the start unfamiliar with this language. Given this circumstance, Al'fred Markovich has been irreplaceable. He's a most cheerful interlocutor and doesn't refuse to join us. Without him ours would be a poor affair. Since we don't like to talk much, we toss out a few words with their help and speak to each other in the simplest short phrases. Aleksandr Gennad'evich prefers to write. He jots down something in his little notebook and Mlle. Dowlen writes her answer. It's hugely amusing, of course.

Having learned that Karl Ivanovich's wife is going with us, Mlle. Dowlen also wanted to go on the expedition. We were understandably overjoyed at having such a sweet companion, but Karl Ivanovich destroyed our flimsy castles by definitively refusing to take her on. She was oddly aggrieved and even cried, and her husband was distressed, but nothing can be done—she'll have to submit and stay behind.

Along with her our companion [J. B.] Landfield was also refused. He arrived with the engineer Rickard and has since become our acquaintance. It turns out he spent time in Russia and speaks very good Russian. We were happy with him because this meant there'd be another person who might be able to show us many things. In addition, he was such a cheerful and sociable character. True, he talked too much about himself, about how he completed two majors in university while also managing to achieve the rank of captain, has been to Russia twice and lived there a long time, and all this by only age 26, yet we somehow believed him and instantly established friendly relations. Karl Ivanovich was very cold toward him and said he couldn't be brought on the expedition. Then Baker and all the Americans besieged Karl Ivanovich and he finally gave in. The next day, a surprisingly courteous and obliging Landfield had already run to the shops, completed all necessary requirements, and managed to get all the necessities for his person.

We've also taken a liking to the engineer Rickard. He's a tall, strapping brunet with friendly, affectionate eyes and is serious, though very courteous towards us. And so, all the expedition's members are present and energetically completing all the requirements for our goals.

Stern recommended his acquaintance, the commissioner Bernstein, to make purchases at our discretion. Bernstein ran around and did some work, though this didn't mean we were completely freed the labor of shopping. We accompanied him and ourselves looked over and selected goods while he simply wrote them down and covered the cost. In this way, though we found him practically unnecessary, here such a procedure is performed by a commissioner, strictly speaking. For his labors, we had to pay him 2 % of the total spent. And this is a very roundabout figure. Stern himself procured the food provisions and spared us this task.

So, we were running to the shops, but Al'fred Markovich was busy preparing the ship for the expedition. He had his work cut out for him. There were no quarters for the workers, so several cabins had to be built on the deck's prow. To do him justice, Al'fred Markovich did everything very well to the extent possible. Across a small expanse he managed to situate a bathroom, photography studio, kitchen, two small cabins, dining room, and two large rooms for the workers, and all this rather comfortably. Then, a steam heater, of which there were none on the vessel, was built, though it is not especially convenient. The passenger cabins were also fitted out. In a word, Al'fred Markovich has worked like an ox during the day and still spends time with us in the evening. It's just unpleasant that all these alterations are being charged to the expedition, and that's several thousand dollars.

Finally, the steamer having been readied, the loading began. This business was done quite carelessly. They brought the supplies straight to the shore and began to stow them in the hold without any system. A certain young man, formerly the ship's baggage foreman, was in charge of this, and how stupidly he did everything had to be seen. From the first day, it became clear that to find anything one has to rummage through the entire hold, and besides, no one knows where anything is. The loading has been finished just before departure.

Baker held a farewell supper for us several days ago and there, where we were dining, yet another gentleman, [John A.] White, arrived at the restaurant, asking to join the expedition. Karl Ivanovich definitely refused but Baker, citing his authority, continued to insist for several hours and Karl Ivanovich finally relented.

In the end, he has become the final member, and our expedition is proving to be fully staffed. Beyond all our expectations, this staff has turned out to be greatly and preponderantly populated in favor of the Americans. We Russians have proved to be as follows: expedition leader Karl Ivanovich and his wife, the doctor [Akif'ëv], Aleksandr Gennad'evich, the old seafarer Lemashevskii in the capacity of reserve captain, should there be such need, and as adviser concerning the sea voyage, and herein may also be included Al'fred Markovich who, despite traveling as an Englishman, belongs in spirit to our Russian party.[11] Our more numerous American companions are, by name: Roberts, Rickard, Dowlen, Stern, Landfield, White, as well as a trio who are neither workers nor assistants: Picard, and [Charles] Scranton [Sr.] and his son [Charles Jr.].[12] Our ship's captain is [Edward] Jahnsen, a short, fat Norwegian, always a bit tipsy, very proud of his 32nd-degree rank in the Masons, and wearing on his vest a Masonic symbol in the shape of a cross. His two assistants, as well as the sailors, stokers, oiler, cooks, and lackeys—this is the ship's crew—are all Americans.[13] Finally, Roberts strenuously insisted on bringing several Chinese. Karl Ivanovich agreed, and Bernstein was persuaded to hire them. Right now, there is a quarantine in San Francisco due to an epidemic in the Chinese Quarter. A fence is blocking this quarter and interaction with it is

---

11. Akif'ëv neglects to include his Ossetian lackey Nikolai in this list, which is telling.

12. Some sources suggest that Picard and White were English. However, they were more likely American. Rickard, as previously noted, was Swiss but living in the States at this time (see Dramatis Personae).

13. Jahnsen's assistants were Kinney and Reiner (see Dramatis Personae).

prohibited. Getting the Chinese wasn't easy; Karl Ivanovich had to telegraph Washington. Permission came from there to take the Chinese for the Russian expedition, but preliminarily to subject them to a thorough disinfecting. For his part, Karl Ivanovich has chosen to bring several Russian workers.[14]

A new acquaintance we had made helped us to hire them.

Sightseeing in the city one day, Aleksandr Gennad'evich and I were seeking out the Russian church. We didn't know where it was, and we turned to a policeman. He very kindly led us to a tramcar and explained to the conductor where we had to transfer to another car. In this way, we reached the church successfully.

The church here is small, quite beautifully appointed, with a very simple though tasteful interior. We arrived during services and stood among the praying crowd.

It's an unusually nice feeling when you attend an Orthodox church abroad. Remembering back, you somehow get the idea that attending church is better than it was during your childhood in Russia. With various foreign words clanging in your ears always and everywhere, it becomes somehow relaxing when you hear familiar prayers in your native tongue, see and talk with people, and don't feel alien. The priest here delivered part of the service in Slavonic and part in English. He was backed by a small, not very harmonious choir. It proved to be a choir of amateurs, many of whom could not speak Russia, though they sang very clearly and grammatically.

We stood with pleasure all through Mass, and after Mass a gentleman approached and introduced himself, asking us not to refuse to have tea with him. He'd clearly seen we didn't look like the Russians living here, and we went with him. Accompanying us was one of his acquaintances, Nepriakhin.

We walked a short way up Green Street and climbed through a grove up a sheer cliff, on top of which a small copse was concealing a small, one-story private home where our new acquaintance, Fedorov, lived. There was a splendid view from there of the city extending to the bay, where sailboats were lightly gliding. The city noise hardly reached us, and the aromatic flowers covering a wooded garden sweetened the air.

"Truly, is there anything better than my little spot?" Fedorov turned to us. "It's just like I'm living in a dacha, and the tramcar and a row of shops are only two steps away! In a word, all city conveniences are at hand, but here, in this little garden, at this altitude, you forget you're in the city! I have a whole farm here, really, I even keep chickens and doves, and all this doesn't run me much. But let's go inside for tea!"

---

14. Vasilii Beliaev, James (Timofei) Sanderson, and Grigorii Shurduk (see Dramatis Personae).

The inside was clean, bright, and so comfortable and quiet after our Palace Hotel's noisy streets and constant crowdedness. We began discussing our journey, and our new acquaintances were astonished by the cost of leasing our ship.

"Really, you've simply been swindled, you could pay less and get ownership of a much better and more suitable vessel for you, and if you want, I'll show you a ship made to order that is far better than your *Samoa*. Yours is really a freight steamer for coastal shipping. It's been transporting timber here. Moreover, it's no wonder you were swindled, since you yourselves didn't purchase the ship other than through commissioners. Well, here, you should know, one hand washes the other more than in other places. And until you familiarize yourself and get used to it, it will be expensive for you. It was just as hard for me. I came here under certain circumstances preventing my living in Russia, and at first, I practically starved. But then I established myself, began learning English, and as you see I don't live bad. I'm a metalsmith and work in one of the factories here. I get 105 dollars a month—here a worker's pay is very good, and though I don't have a choice, I'll stay forever.[15] It's very expensive for newcomers here, you know. You, for example, are staying at the Palace Hotel, a hotel specially for the wealthy. Of course, it's costing you no less than 10 dollars a day, but if you'd turned to people familiar with living conditions here, you could have gotten a room for 20 dollars a month, perfectly nice, and a satisfying meal for four times less than at your hotel. There, as you say, they add a dollar and a half corkage fee to the price of your wine, but that's quite strange because wine is cheap here in California and I have wine on the table every day, though I buy it by the gallons. No, I've gotten used to living well here, though of course the homeland's always calling. I've been living here a long time already, got married here, had children, and have generally set myself up pretty well. Few Russians live here, but if they're capable and want to work they can do pretty well."

Having finished tea, we went to our place after having asked Fedorov to send us any Russians who wanted to leave with us. Three such Russians were found.[16]

Everything's nearly ready and we've begun organizing our cabins. Our steamer's small, all of 150 feet long with a width of 32 and a draft of 12 ft. The cabins are small and therefore get crowded. The cabin companions gathered on deck and entered through the seven cabins' doorways, with the engine and kitchen occupying the upper level's rear structure. Above this level is

---

15. This hints that Fedorov was probably a political dissident back in Russia.

16. On San Francisco's early Russian community, see Terence Emmons, *Alleged Sex and Threatened Violence: Doctor Russel, Bishop Vladimir, and the Russians in San Francisco, 1887–1892* (Stanford, CA: Stanford University Press, 1997).

a second level surrounded by the bridge, and above the bridge the sloops and a place for the watch have been situated. In the front of the second level is the helm, a cabin for the assistant captains, the captain's cabin, and four more cabins. One, formerly the smoking room, has been assigned to Karl Ivanovich, the second is to be used as the pharmacy, where I've been assigned, Lemashevskii is occupying the cabin next to me, and the last, rather small one will be occupied by Al'fred Markovich and Rickard.

It's more crowded below. Aleksandr Gennad'evich has gotten his own cabin, as has Stern, but Roberts is sharing with Landfield, and White with Dowlen. The engineers are occupying a single cabin, the Stuarts another, and Nikolai and three engineers the final one.[17] There is absolutely no space left.

Hence, all is ready and tomorrow we'll board the steamer and settle in for good and may leave tomorrow.

---

17. This is Akif'ëv's only reference to the Stuarts; later, he makes another, unflattering reference to a Stuart in the singular. We learn nothing more about these people.

CHAPTER 3

In the Great Ocean

Aleutian Islands—Bering Sea

28 May/10 June. Here we are at sea. We left San Francisco at 12:00 o'clock last night, and did so quite uniquely. Yesterday, we dined for the last time in the city, and while there I learned from Al'fred Markovich we would be leaving that night.

"But what about Roberts and Dowlen?" I asked. "Don't you know they're not on the steamer?"

"It's a big secret; as a matter of fact, yesterday the police detained Roberts and Dowlen and forbade them from leaving, so we have to escape today."

"Let's go say goodbye to Mlle. Dowlen."

We went to a room in some restaurant where everything was already arranged, and found Dowlen and his wife with the senior mechanic, Kinney. Everyone was unusually secretive. We drank a farewell bottle of wine, said goodbye, and parted.

At midnight the steamer was completely dark, without a single light. Only Bernstein and Stern's brother were on the dock. The goodbyes were brief, and the ropes were cast off. Without a whistle, silently, just like a sneak, the *Samoa* slipped from shore and headed to sea.

For an hour, we slowly, slowly passed by the boats in the road. From shore, a small black dot appeared in the dark. The steamer came to a stop. The dot increased and grew until finally, we were looking at a boat gliding quickly and noiselessly toward us. It moored alongside, and Roberts, Dowlen, and Scranton and son came aboard. Just as quickly, the boat left. The *Samoa* blew a whistle, immediately lit everything up, and went full steam ahead to sea. How laughingly that whistle sounded, and what an impudent note it sounded!

I asked Al'fred Markovich what was going on. It turns out Dowlen didn't fulfill some contract, Scranton was arrested for some petty debts, and Roberts for large ones. In America, there's a law that if an insolvent debtor goes to another state the matter carries over to that state's police, and so the business drags on and such opportunistic escapes are not rare. Be that as it may, such a solution was not to our liking. We're going off to a job, off to a harsh, remote land, so why hide, why be ashamed and leave in secret, like thieves?

All the same, these Robertses, Dowlens, and the rest are cunning people. A day before, they spread a rumor that the *Samoa* would be leaving at 12 noon on Sunday the 10th; they didn't show up for the steamer but hid out somewhere, and during the evening they quietly left San Francisco, hired a boat, and taking advantage of the darkness and unnoticed by anyone, left to meet the steamer, which had previously been apprised of the appointed time. Arriving at the dock the next morning, the police will have found an empty spot.

Following this, as the steamer proceeded through the sea, we drank tea to a fortuitous journey, and the expedition commenced. But because we're divided among separate cabins, the celebration was not cheerful.

So, we're on the ocean again, the only difference being that earlier we were in the Atlantic Ocean aboard the enormous ship *Campania* and now we're in the Great Ocean aboard the small ship *Samoa*. If the former slid down a wave many times its height like it was a mountain, it can only be imagined what will happen to the *Samoa* during a storm, to this pygmy in comparison to the *Campania*. What will happen if 10 inches of water suddenly cover our *Samoa*'s deck? Whereas the *Campania* could surmount the swell, it'll be interesting to see how the *Samoa* fares.

Today the ocean is like a mirror, but even so, the steamer is rocking. California's foggy shoreline remains visible to the right. The air is damp and warm. In a few hours we will say, "Farewell, California, farewell, land of gold, sun, and warmth," and in greeting say, "Harsh North, we approach you as visitors!"

29 May/11 June. Cloudy, but warm. We're proceeding slowly northwest. The sails are unfurled. The boredom aboard the steamer is terrible. It's crowded, there's nowhere to walk. The whole deck is piled with sacks of coal up to the bridge. Aleksandr Gennad'evich and many others have been laid up with seasickness. All last night I was sick as well, and though I'm better today I'm not entirely well. I give the sick a kind of mint liqueur and everything the captain advises and promotes, though I don't see much use in them. Walking around is impossible and the jolting is so spasmodic it's hard to stay on your feet, so I lie on the bridge and read, though our Americans play dice lying on the floor. Gennad'ich has also been lying down all day, sprawled on deck, enamored of the little sloop.

3/16 June. The Great Ocean hasn't pampered us, we're being rocked and barely allowed a sigh. The rocking lessened only on the morning of the 13th, but the rest of the time, as a matter of course, we rock along the keel and then to the side, sometimes quite heavily. There are swings as much as 65° to either side. Everyone's utterly put out, and it's impossible to undertake anything seriously. We lie on deck and read all day, but this is quite boring. The sky is overcast for the most part, but when the wind speeds the clouds along, we relax for several hours marveling over the wondrous picture. Rays of sunshine pour down from a clear sky onto mighty, deep blue waves. The wind blows, the ocean calms, the waves settle, and, like a silvery coat of mail, the hero-ocean's chest gleams.

The water beneath the steamer itself is of an azure color, as in the Mediterranean Sea. A flock of albatross is flying behind us. They're not at all afraid of the steamer, and whenever the cook tosses something out, a whole bunch of them descends with cries and practically snatches it from his hands. The Americans amuse themselves by tying pieces of meat to both ends of a rope and throwing it in the water. In a moment, a flock of albatross appears and goes swiftly after the meat, stealing it from one another. But when a pair contrives to swallow it, they, in a sense, prove to be on the same rope. Then each for its part begins pulling the other toward it, until the stronger yanks the piece from the stomach of the weaker.

Albatross are not our only traveling companions, as there are whales moving in the same direction. Waterspouts break the surface here and there, sometimes so close to the gunwale you can hear them snorting. An entire pod of 6 or 7 individuals will often catch up to our ship and cross its path, turning themselves over and showing off their gigantic split tails. Like us, they're going to the Bering Sea and onward through the strait to the Arctic Ocean.

It's quiet on the steamer, and for days the Americans have been playing cards or dice and drinking in the evenings. Many are sick. Picard's seasickness has gotten very serious. He's been vomiting endlessly and now he's vomiting blood; I'm giving him codeine with bismuth, and this seems to have helped.

Our Russian hires have been assigned as follows. First, Vasilii Beliaev, a former sailor, is serving us in the capacity of day-watch. James Sanderson, whose Russian name is Timofei, is working as a sailor, and the third, Grigorii Shurduk, is helping Al'fred Markovich straighten out the cutter Karl Ivanovich bought in San Francisco. Because our ship has few sailors, Timofei and Grigorii are working a lot. The coal bins are already empty and need refilling, so both are using sacks to transfer to the bins the coal filling the prow's hold. I was in their proximity, and they were complaining about having to work all day whereas the Chinese do absolutely nothing. I told Karl Ivanovich about this, and he ordered the Chinese to get on deck and work. The Chinese refused, saying they were all ill or seasick. I visited them and saw that half were lying down, but the other half were completely fine and walking around the cabin. There was a strong smell of opium in the cabin. The eyes of those lying down were utterly stupefied and glassy and it was obvious they'd been smoking.

I said to the senior Chinese: "Here they are healthy, and you're healthy, so get to work."

"No," he answered. "I'm the leader, so I'm not going to work, and they're not obliged to work aboard ship but will work when we reach dry land."

A threat to withhold their rations failed to materialize. Then the captain promised to douse them with water from the pump if they didn't get to work. After this, several men actually went out, but how they worked! Two of them would carry the sacks of 3 to 4 poods,[1] whereas Timofei, as if for fun, would toss a sack on his shoulders and easily walk it by himself between the sacks on the rolling deck. Seeing how they worked was offensive. They can be contrasted to those Manchurians who work for us building the railroad in Transbaikalia. Recalling the half-naked Manchurians' tall, strong, bronze physiques and comparing them to our small, ugly, weak Chinese, you think: "Why did we take this horde of good-for-nothing, capricious, and obstinate monkeys with us?" And their pay! 60 dollars a month in ready maintenance, and in case any of them dies, delivery of the body to San Francisco. You see, they have a law that a Chinese has to be buried in China. I can imagine how nice and comfortable it will be next to a decaying corpse on our already packed ship.

Last night, there was a strange rocking. No matter how hard I tried I couldn't fall asleep. I was banging my forehead on the wall, being tossed from side to side, and barely staying in my berth. The vials on my shelves were bouncing in all directions, resounding in various tones. Mortars, glasses, and books were traversing the table. Clothes were flying off the wall into the opposite corner. The whole steamer was creaking and shaking. So many times I had to get up and turn on the electric light so I could gather the items thrown in every direction. Only toward morning was I able to fall asleep. There was the same rocking during the morning, and a thick ashen fog surrounded everything. Gray waves

---

1. Three to four poods is equivalent to 108 to 144 pounds.

smashed against the gunwales and sometimes vaulted onto the deck. It was damp and cold (10° C). We crawled into the hold, but it wasn't safe there. Boxes piled on top of one another were suddenly shifting, and we, not wanting to risk being crushed and having removed several items, went on deck. Of course the Americans, seeing we'd removed some items, immediately rushed to sort things out according to ownership but really couldn't put their finger on anything.

It's evening now. We're sitting in my cabin. I'm writing my diary, Aleksandr Gennad'ich is humming quietly and playing something on guitar, and Nikolai is sitting on the floor; Vasilii is leaning against the doorjamb and we're both silently listening to his mellow, dolorous singing, but the wind from outside hisses monotonously and cold waves splash uniformly against the sides.

7/20 June. The Aleutian Islands. There was a knock on my door and the voice of Aleksandr Gennad'ich: "Get up, land's in sight!" Jumping up and cursing, I hurriedly began dressing. And the reason to hurry was that it was less than 10 degrees in my cabin and getting out from underneath the warm blanket was not very pleasant.

Ascending to the bridge, I was astonished by the splendidly majestic scene opening before our eyes. To the left, the ocean was muffled in thick fog, but to the right, the clouds were breaking, and bright shafts of sunlight poured through. And up ahead, a great pile of mountains, lifeless, gloomy, touched only in places by pale green spots of vegetation, was rising perpendicularly above the water. Their peaks, covered in snow, gleamed dazzlingly in the sun, and further beyond them still taller peaks ascended into the clouds, only their lower slopes mottled by areas of black and white. This was Cape Morton.[2]

A flock of birds was flying above our heads very low beneath the clouds, but no life was visible onshore.

Having stopped to the right of Cape Morton and proceeding several miles in a straight direction, we turned sharply west and entered the bay. There an argument erupted between Jahnsen and Lemashevskii about where to stop: Dutch Harbor or Unalaska. Lemashevskii was for the latter, and Karl Ivanovich took his side; in such way, we bypassed Dutch Harbor. Several ships were anchored there—three American warships, one British one, and two or three merchantmen. A coaling station is there. At this time of year, a given steamer can load there but cannot turn around and go to Cape Nome because of the ice pack it will encounter.

---

2. A "Cape Morton" (or "Moreton") does not appear in a Google search for this part of the world. Nonetheless, additional information below indicates the steamer was entering Unalaska Bay, on the island of the same name.

In Unalaska, separated from Dutch Harbor by just a small promontory, the first thing that caught our eyes was a wooden Russian church with a green cupola, small and rather pretty, or so it seemed to me, perhaps because I immediately caught a whiff of home. Approaching to just a few dozen *sazheni* from the coast, the steamer halted and released its sloop to the shore. Having returned from shore, Reiner and Al'fred Markovich pointed us toward the wharf, and several minutes later we were throwing out the gangplank.

After ten days of uninterrupted rocking we were finally seeing solid soil. No wonder our first desire was to go on land, and Gennad'ich and I made straight for the village.

Unalaska is a very small village of thirty or forty small buildings. They're spread along the same bay's coast, beneath mountains from which they're separated by a stream. Twenty of these little structures have been built exactly the same, in the manner of temporary barracks, and are painted one and the same dark red color. Each is numbered. These are the dwellings of Aleuts previously under contract to the Alaska Commercial Company, but now they work for it in return for what the company offers as housing.[3] The little buildings' interiors seemed tidy, and muslin curtains and flowered windows were visible everywhere. They were very lightly constructed, evidently from just thin planks. One has to be greatly habituated to winter to live in such light structures; but they're used to it because nature is sufficiently harsh here: despite it being the 20th of June, the mountains are all still covered in snow.

We saw practically in the middle of the village a school, on which was nailed a placard signed by the Rev. John Veniaminoff, and further on the church.[4] Behind the village, on the bay's left corner, could be seen a cemetery, toward which we went.

Along the way we encountered several Aleut boys with swarthy faces, rigid black hair, and large, oblong black eyes, and we were very surprised when, after

---

3. In 1868, following Alaska's sale to the United States, the Russian-American Company (RAC) (chartered by the Russian government in 1799 and similar in function to the British East India Company) was sold to an American conglomerate, which renamed the RAC the ACC.

4. Also known as Saint Innokentii, Veniaminov was born Ivan Evseevich Popov in 1797, in Siberia. Arriving in Alaska in 1823, he was the first cleric to proselytize Russian Orthodoxy in the Americas. It is curious that Akif'ëv uses his anglicized name and spells it using Latin letters, and, moreover, gives his honorific as "Rev[erend]." On Russian Orthodoxy's early history in America, see David Nordlander, "Innokentii Veniaminov and the Expansion of Orthodoxy in Russian America," *Pacific Historical Review* 64, no. 1 (1995): 19–36; Veselin Kesich, "The Orthodox Church in America," *Russian Review* 20, no. 3 (1961): 185–93; Jesse D. Murray, "Together and Apart: The Russian Orthodox Church, the Russian Empire, and Orthodox Missionaries in Alaska, 1794–1917," *Russian History* 40, no. 1 (2013): 91–110; and Andrei Znamenski, *Shamanism and Christianity: Native Encounters with Russian Orthodox Missions in Siberia and Alaska, 1820–1917* (Westport, CT: Greenwood Press, 1999).

FIGURE 3.1.   I. N. Akif'ëv

having evened up, they greeted us with "Hello" in Russian. Really, though the Aleutian Islands were given to the United States in '67, a Russian influence nevertheless remains.

On the way to the cemetery, we saw in a meadow a bunch of violets completely without scent, albeit large and dark purple. A broad variety of flowers was in the cemetery: forget-me-nots, pansies, swathes of buttercup, and even more that I'd never seen before. Gennad'ich and I ran along the hill gathering flowers like children, and after a little while we were holding large bouquets. It was so nice to stretch out our legs and to feel beneath them hard soil and not the pitching deck, and to breathe pure mountain air perfumed by the smell of grass and flowers. Having run about, we reclined on the green grass with pleasure after the rigid, narrow steamer bunks.

The cemetery was located on a hill and divided between the Orthodox and non-Orthodox. Barbed wire fenced off the Orthodox sector, Catholics and Protestants lay outside the fence. Some graves bore a trellis decorated with flowers. Small marble markers could be seen above the sailors interred there.

Having returned from the cemetery, we were met by Grigorii, who said the local priest was sick and had asked me to attend to him. I went, and found Karl Ivanovich and Lemashevskii talking with the sick man.

The priest, Rysev by name, was an old man of 72, tall, with a kind, sickly face. He's devoted his entire life to his mission, and spent all of it on these harsh islands; he's buried his own son here and will himself die here. Seeing the unpretentious circumstances of his little domicile, so familiar to this harsh, barren land's half-savage people, you felt such profound respect for this man, this modest, undemanding bearer of the Orthodox ideal, and for what a deep rift there is that separates ours from those Catholic and American missionaries who live in Korea and China, who imbue their missionary activity with political agitation and will, within an hour, take advantage of any pretty thing they happen to see. But here was pure simplicity. His daughter, a native of the Aleutians and also the wife of a missionary, was there as well, along with her beautiful little daughter, a little girl of 10 holding a small baby in her arms, whose swarthy-colored face and large black eyes bespoke a creole. Quite curiously, everyone called him papa and spoke Russian well and grammatically correctly, except only that he himself sometimes let slip a "Yes" in English.

Having examined him and the baby, I promised to get some medicine and left for the ship. The sailors there had been catching fish very successfully. They'd pulled in around ten large fish, nearly an *arshin* long. Consequently, we were going to have fried fish for supper. After having gotten sick and tired of canned food, not bad.

Toward evening, Gennad'ich, Al'fred Markovich, and I headed over the hill for Dutch Harbor. Absolutely nothing of interest was there. Onshore were two or three little buildings, a shop, and a coal depot; nearby was a bog. On the way back, we were surprised that all the surrounding grass had been burned. Learning why was interesting. The matter goes as follows: American prospectors consumed with gold fever travel through here on their way to Alaska; to them, gold seems to be everywhere, and widespread among them is an opinion that smoke can reveal where gold is in the ground, so they've torched all the surrounding grass.

8/21 June. I awoke this morning and nearly screamed from the cold. It was minus 7 degrees Réaumur in my cabin.[5] The sun was shining clearly, the water

---

5. The equivalent of 7° Réaumur is 48° F, or 9° C.

FIGURE 3.2.   A. G. Miagkov

in the bay completely green, and snowy peaks were gleaming like white teeth around the backdrop of the pale blue sky. During the morning I prepared the priest's medication and brought it to him, instructing him how to take it. He was very grateful and asked me to take two sealskins as keepsakes. The skins were gray, unpainted, and soft as velvet. I was inclined to take them, but knowing how valuable they are here, I refused. We'd priced them in the local shop and were told they sold for 15 dollars a hide.

After lunch, Gennad'ich and I headed for the mountains and found ourselves a summit from which extended a beautiful view of the surroundings. But we mistook the road and went to the right up a rather large hill from which we had to descend and got very tired. It was getting hot, and we were dressed in warm jackets; so we climbed only as far as the snow line, a hundred *sazheni* up, and lay down to rest. The view from there was spectacular. Below, spreading like the palm of a hand, was Unalaska, and visible farther away was Dutch Harbor and the ships in the bay, and still farther the sea, lost in the foggy distance. The fantastic, snowy forms of peaks surrounded us wherever you looked. Over there would be some crater of a dead, conical volcano whose enormous funnel went deep into the mountain, and over there, stretching into the clouds, was a peak as white as a head of sugar, while there was a dark precipice conjoining the sea to the clouds. Barren wilderness surrounded us. The quiet was deathly and strange, broken only occasionally by the call of a crow and the screams of eagles flying above us. I spread out my arms and lay motionlessly, gazing upward. A large, soaring bald eagle noticed me and, making circles, began to descend. As soon as he was close, I pulled out my revolver, fired a bullet, but missed. We both shot several more times but without success, then left for the shore. Along the way, we came upon a reservoir whose water supplies Unalaska's inhabitants. This is a lake quite high in the mountains, with splendid water and narrow pipes that feed Unalaska below.

Here, we've sold for 10 dollars apiece the sheep we earlier had on ship, and we've purchased an entire ox for 25 cents a pound—i.e., 50 kopeks. As you can see, not a bad price. The several Russians living on the island gave us a send-off. We left Unalaska at 6 o'clock this morning and had only just entered the open sea when we began getting tossed around like a ball. A bottle of milk in my cabin slammed into the washbasin, smashing it to pieces. By the sea's grace, I'm now without a washbasin. Little by little, everyone's starting to collapse: best to refrain from sin.

9/22 June. In the Bering Sea. The rocking's powerful, the cold is blowing from the northwest, and we're proceeding at no more than six knots an hour. The water is already just 4.5 degrees: it's quite evident we're in the northern sea. Nearly everyone's lying in their cabins, the Chinese are as well and won't go on deck, though their quarters in the prow are very stuffy and foul. You're presented with a general filthiness that is of the particular stench that emanates from a Chinese, a stench of opium and traces of seasickness, and you can know this because of their cabin's air quality. One day Reiner (name of the senior assistant captain) dropped in on their cabin and immediately vomited. With great effort, we got them to bring all their bedding on deck so it could be aired out. Meanwhile, we managed to air out and wash their quarters with carbolic acid. In such way, they became less offensive.

In the evening Gennad'ich and Vasilii came to my cabin as usual. Gennad'ich grabbed his guitar and played doleful Russian folk motifs. Vasilii, for want of space, sat in the doorway and listened, holding his head in his hands. His face, lean, with a sharp nose, impoverished blond vegetation, and narrow colorless eyes, revealed avid attention, yet at the same time it was clear that under the impression of the folk motifs, his thoughts were being carried far away. The song's plaintive murmur led him to pour and pour everything out. He'd blinked his red eyes and shaken his head to adjust his cap several times already, but finally could not hold back and said, in neither a shout nor a sigh, "This song is so good! It really takes me back to Russia! Why am I wasting my time here? I'll earn money from you and leave for Petersburg."

"Have you really been away from home long?"

"I was home two years ago but haven't been in Petersburg for a long time. I want to live there. I served seven years in the navy, and after my service I entered Count G.'s employ as a lackey. I served him for a full year. There wasn't a position any better, and the count was a painfully stupid character, first cursing me then praising me for no reason. I tossed it—that means, I left that position for home. I lived three whole years there and worked as a border guard for Viatka Province.[6] It got boring, and I thought to myself: 'Why am I sitting here? I need to see the world and get a better salary.' I'll have you know I went straight to Tsaritsyn[7] and found a good job working on a hotel buffet. I know this business, and for the past three years at sea I've also been on the buffet. My salary was good and so was my income. But I didn't stay there. After 4 months I went to Astrakhan and joined a steamer as the assistant buffet director. It would've been good there, of course, but I argued with a buffet attendant who was the captain's lover, so they sacked me. Luckily, the machinist gave me a recommendation and I went again to Astrakhan, to work as a waiter on a schooner in the Caspian Sea. I could've served there awhile as well, but somehow couldn't get along, and left for Uzun-Ada.[8] 'Where'll I disappear to now?' I thought. However, I found a position on the railroad as a rifleman, for 14 rubles. In a short time the sector commander took me on as his lackey and gave me a 5-ruble raise. From Uzun-Ada I began traveling the entire length of the Transcaspian Railroad, then under construction. I served in Samarkand, Margilan, and Kokand. After almost three years, the construction was finished, and I went home to Kostroma Province.[9]

"I'll have you know there wasn't much of a life at home: nothing was happening, and I was getting used to being lazy. So, that winter before Christmas, I got ready to go to Nizhnii.[10] I found a traveling companion for two-and-a-

---

6. Exiles escaping west from Siberia often passed through Viatka Province. Vasilii would have been assigned to catch them.

7. Tsaritsyn was later renamed Stalingrad, and today is called Volgograd.

8. Uzun-Ada was a port on the Caspian Sea.

9. Kostroma Province was just west of Viatka Province.

10. Nizhnii Novgorod was the capital of Nizhegorod Province, south of Kostroma.

half rubles, and we set off. When I arrived, I tried to find a position, of course, but there were no jobs. They said there was a spot in a religious seminary. I went and asked. They said there was a position for a watchman at a salary of five rubles a month. 'If you want,' they said, 'come in two weeks and we'll take you on.' I started thinking: 'This is really insulting, I've worked on steamers and in a hotel and earned up to 60 rubles, but here's five rubles and I need to wait two weeks. No, I won't have it.' So I didn't go for it. I began asking around more. Word was a tannery in the village of Bogorodskoe needed people. I went on foot, all forty versts. I got there, asked for work. 'No,' they said, 'we don't need anyone.' Where to go? I go back to Nizhnii. I'd gotten to know a fellow there, also sitting around with neither work nor a farthing. We got to talking, thought about it, and decided to go to Astrakhan for the winter, to be there when spring came. We got ready and left for the coast. We got as far as Lyskovo, but something snapped between us, and I went on alone to Kniaginin and to Penza.[11] But in Penza, I decided to go to Astrakhan, to go, I'm thinking, on the Don. Of course, I asked around, and away I went. I walked the 800 versts through Ust'-Medvedevsk Station and Novocherkassk and reached Rostov.[12] No jobs. 'Well,' I told myself, 'I've walked so far there's no turning back, so I better keep going.' I went to Taganrog, Mariupol, Kerch— no jobs anywhere. 'Well,' I'm thinking, 'I'll go to Odessa.' I got to the wharf and found a steamer going to Nikolaev.[13] I go to the assistant captain: 'Your excellency,' I said, 'take me to Nikolaev for work.' 'Very well,' he said, 'come aboard.' I reached Nikolaev and there the assistant gave me a recommendation for another steamer bound for Odessa. Once again, there was nothing in Odessa. I got work not far from Odessa, building a house near Ochakov, and heard people were needed in Vladivostok to build the Chinese railroad. I returned to Odessa and arranged with the Volunteer Fleet to go to Vladivostok for work. Nowhere did I pay for my travel, I'll have you know, though had I been able to work, I would've paid. That's how I got to Vladivostok and worked on the railroad[14] for a rather long time as a senior laborer, but some people stirred me up: 'We're going to find gold,' they said. 'We hear there's lots here.' I'm thinking and thinking, and decide I want to strike it rich. I tossed my job, and a trio of us left. We wandered the forest for three weeks, one time some

11. Kniaginin and Penza were, respectively, to the east and south of Nizhegorod Province, which means that Vasilii was practically going in circles.

12. Vasilii was now far to the south, where the Don River enters the Black Sea.

13. Nikolaev is modern-day Mykolaiv, Ukraine.

14. Built by Russia and extending into the heart of Manchuria, the Chinese Eastern Railroad (CER) was actually far west of Vladivostok. If Vasilii did indeed work on the CER, he must have traveled inland from the city. By contrast, the railroad local to Vladivostok was the Ussuri Railroad, the Trans-Siberian Railroad's easternmost extension. Vasilii might be confusing it with the CER.

Chinese nearly murdered us, and we got emaciated and insane but didn't find a trace of gold. 'Well,' I tell my comrades, 'we're going round and round to no avail. You can keep searching, but I'm going to Nagasaki and won't be with you.' I wanted to get to America, because it was said you could earn a lot of money there. I can say that for as much as I've traveled, I haven't had much luck from it. Of course, I went to Nagasaki for work and was staying in a certain Jewess's apartment. We got to talking. She knew I wanted to get to America and told me: 'To hell with that America,' she said, 'I lost four thousand rubles there. You shouldn't go,' she said. 'But if I've got nothing, what can I lose?' I thought. Well, of course I went to a steamer and asked to work for passage to San Francisco. "No,' they said, 'you can't.' How was I going to get there? If it was like they said, I'd have to travel as a stowaway. So I did. I went aboard the steamer two hours before departure. As it was getting ready to leave, I went into the water closet and sat there until we left.

"I come out, but run into the assistant I'd asked about passage. 'How'd you get here?' he says. 'I told you it wasn't allowed, y'know!' 'Do what you want,' I told him. 'Well, damn you,' he says, 'get moving, you'll work in the coal bin.' Another stowaway was on our steamer, an American they made fix the sails. We were signed on as sailors, which meant we were supposed to get paid 20 rubles a month. When we got to San Francisco, they ordered us to sign a receipt. 'What for?' We had to sign, so we signed, but didn't get our money, of course, and the captain must have gotten it. I went to the consul in San Francisco. 'Your Excellency, I came here to work, of course, so don't stint on the advice.' 'You don't know how to speak the local language,' he says, 'and without the language you'll be bad off here. My advice for you is to turn around and leave.' The captain of the ship I'd arrived in was there as well. 'I'll agree to take him on the return run,' he said. I'm thinking, 'I shouldn't agree, I'll work, and you'll end up taking my salary.' I was about to agree, but then I thought: 'What did I come to America for? No, if I've already arrived, I should see how people live here, and tea will keep me from feeling hungry.' So, I didn't visit the consul anymore. I found there was a lot of Russians in San Francisco. They began to treat me and take me to the theater. But I told them: 'Brothers, I don't have any money, y'know.' Well, they stopped taking me to the theater. I began wandering the city looking for work. What money I had, I spent it all. 'No,' I'm thinking to myself, 'I'll go to the village.' I walked a long time, kept asking if there were Russians, and finally found them. They turned out to be Jews. 'I came here to work,' I said, 'but I don't know the local language.' 'It's hard without the language here,' they said, 'but we can help you. Work for us.' I started working for them. I worked three weeks, they fed me, but said nothing about my pay. 'What a nuisance,' I thought, 'serving without pay.' I asked for compensation. They gave me a dollar and a half for three weeks. 'You're not paying much'—but what could you do? I took the dollar and a half and went wherever my gaze took me. I'm following the road, asking: '*Work have Jou?*' meaning, 'Do you have any work?' '*No work.*' Finally, in a certain spot in the mines (in English, they pronounce it 'maynz'), they said: '*Yes, we have.*' 'How much per day?' I asked. 'Two and a half dollars.' 'Very well,' I said. 'I agree.' After lunch, I began the job. They gave me a wheelbarrow and stuck me in a spot where you had to bend in half and couldn't turn around. 'Well, it's a job!' I was breaking my back, bending over to work, felt like a smithy hammer was banging in my head, and I had no air. However, I worked till evening. '*Very good man,*' they said. The next day, they gave me an easier job. So I worked for over a month. That meant I couldn't sit around too long, and it got annoying. 'I'll go vagging again,' I thought. I asked for my pay. Took it and left. Of course, first thing I did was go to a pub, a saloon, according to them; well, I had one glass after another of whiskey then beer and after three days didn't have a farthing. Had to work again. I went searching and found something for two and a quarter dollars. 'Alright,' I thought, 'but I have to walk forty miles there.' Well, luckily, I got a traveling companion, some American who fed me, and we drank so that all his money was gone as well. We get to the job, and of course it was groundwork and some mines needed draining. Well, there was nothing for it, and I worked for a month; again, it became a nuisance. I left for San Francisco, wandered about, but no jobs there. But I suddenly hear this expedition to Siberia needs people. 'Thank God,' I thought, 'America's a real nuisance, people here don't believe in anything and worship the devil and there's nothing to eat.' So, I came to you."

We had been sitting listening in silence to our Vasilii's odyssey. And I was thinking, in another time under different circumstances he probably could have been a Ermak or Semën Dezhnëv.[15] Our predecessors really are distinguished by a fervor for travel: the Ushkuiniks and the Zaporozhians, lovers of adventure as well, battled Sibir alongside Ermak.[16] But time has passed and there's no place for adventures. Why am I myself meandering to the White World? Really, is it not a similar desire to see and do new things?

The wind whistles and howls outside my cabin door; sheets of cold rain pound on the window and dust whorls across the deck. The steamer creaks and groans; it's rocking. Fog surrounds us; the Bering Sea is greeting us coldly and unwelcomingly.

---

15. The Cossack Ermak Timofeev was the sixteenth-century "Conqueror of Siberia"; Dezhnëv navigated the Bering Strait in the mid-seventeenth century, eighty years before Vitus Bering.

16. Ushkuiniks were Russian river pirates; the Zaporozhians were a host of southern Cossacks. Ermak attacked the Sibir Khanate of western Siberia, though Akif'ëv is mistaken about his military force, which consisted of Don Cossacks and European mercenaries rather than Zaporozhians.

11/24 June. Starting yesterday morning, the sea spoiled us with beautiful weather. I woke early and didn't feel any rocking. I opened my cabin door, and it was simply freezing: the ocean was pale blue with a shade of gray and smooth as a mirror. Wispy transparent clouds were sailing high, high in a blue sky and, on the horizon, the peaks of St. Matthew's Island were glistening brightly in the clear light of the sun. It was utterly peaceful, the air crisp and clean. It even seemed to have warmed up, though it was all of 8° R in the shade. Everyone grew cheerful and boisterous; laughter and joking were heard. But this didn't last long. In the evening, a fog descended from the northwest, a cold wind blew, and the rocking resumed. We once again lay in our favorite spot on the bridge and passed the time unconcernedly until 11 o'clock. It was so bright we could read easily. Gennad'ich allowed himself to bring his bed sack onto the bridge and laid it down. He was cold and oddly clammy. I recommended he go to his cabin, but he didn't want to, and only the temptation of eating pineapples in the buffet enticed him to go below after midnight.

12/25 June. There's been a thick, milky fog, raw and cold, and a water temperature of only 4–9° R. We've been proceeding quietly, blowing the whistle now and then. Just before lunch, we heard what seemed an echo of our whistles. What did it mean? Was the shore really nearby? It was dangerous to go farther because you couldn't see past 10 *sazheni* and we might strike shore. We dropped anchor and put the sloop in the water. Reiner and some sailors got in, accompanied by Gennad'ich. Everyone was so sure the coast was near that they didn't bring fresh water or bread. Charting a course north, the sloop cast off and disappeared into the fog. The Americans sat behind their cards, but I went to my cabin to write my diary under the plaintive wail of the siren. This desperate, piercing wail sometimes simply seizes one's soul.

The sloop returned after several hours. They went so far they no longer heard the siren, but couldn't find the coast and encountered some shallow floating ice and turned back, afraid of being forced off course.

The night is completely white, but there isn't really a night and it's only as if it gets overcast for a little bit. We're sitting in Gennad'ich's cabin, sucking down wine on the supposition that we're celebrating having entered Russian waters. Since the fog hasn't broken and the captain apparently wants to play cards, we're anchoring for the night.

CHAPTER 4

On the Chukotka Peninsula Coast

Providence Bay—We find the suspect steamer *Progress*—Meeting the Chukchi—A graveyard—A Chukchi village—Yurts—We cross to Cape Chaplin—The village of Unyyn—Chukchi clothing—Chukchi hunting— Disorder on the ship—The expedition's spiritual outline—Our captain—At Cape Novosil'tsev—Unanticipated encounter with gold prospectors—Miagkov remains onshore near the Marich River—We go to Alaska

13/26 June. Providence Bay.[1] The morning fog has become even thicker. Aboard the steamer, it was rumored that the captain would himself launch our large whaleboat and set off into the foggy distance to seek the promised land. I got absolutely embarrassed and probably even blushed. I'd really always considered him to be a fully capable fellow. . . . He'd been sitting for entire days at the card table, but it suddenly turned out this wasn't his sole virtue. I was watching them lower and equip the whaleboat. "Well," I thought, "they'll be leaving now": I look closer, and they're raising and lowering the mast again. The junior assistant and two sailors got into the boat; now, surely, the captain would feel guilty. Suddenly, I felt our steamer setting off. What was the story? I looked and saw the captain standing on the bridge surrounded by a whole mob of Americans, and I went up.

---

1. Providence Bay is on the southwest tip of the Chukotka Peninsula.

FIGURE 4.1. Providence Bay

The captain was blowing one siren whistle after and another and we could hear an echo immediately after the whistle. "The coast is near," said the captain. I responded, "Absolutely so, according to the map, but this echo has nothing to do with it." A terrible argument erupted next. All the Americans insisted the echo was coming from shore, off the mountains. I asserted this echo was coming off the cloud of fog that encircled us. I indicated to them the speed with which the sound was disseminating, and said that if the immediately reflected echo were coming from a promontory it would have to be extremely close. "Well, there's science," they answered me, "and there's real life, and if the captain says the echo's coming from shore, that means it is." So we failed to reconcile our views. Two hours later, ice appeared, and behind it, in the fog on the horizon, some kind of ribbon that after some time proved to be mountains. The Americans were jubilant and shouting: "Practice won out over science!" and completely overlooked that during the night we'd been twenty miles offshore.

The closer we approached the coast the clearer stood the black groupings of enormous barren mountains covered with white patinas of snow. These gloomy stern mountains, the fog, the gray sea covered by floating ice, and the penetrating cold and damp created a rather unique, depressing impression. We seemed to have ended up in some unknown storybook world.

Proceeding a little to the west, the steamer turned straight north, and we sailed into the bay, slowing pushing between the ice. Everyone gathered on the bridge and looked through binoculars at the promised land, as the assistant captain stood on the prow and directed the ship. From time to time we could feel the weak ice cracking and crunching. Suddenly, we were just able to notice a rising puff of smoke far off in the distance. What was it? Habitation or the steamer *Iakut*, which should be delivering our laborers here? Each gazed with rapt attention and reported what he saw. To one, there seemed to be two steamers, to another, even three. Then, on the water beneath the smoke, a black strip with dark spots appeared. We determined these to be Chukchi yurts.

As the steamer moved slowly along, I wished it would go full steam so we could sooner begin our activity.

Then, we were able to see a steamer behind the lower spit on which those several yurts were located, and sailing toward us from shore was a sloop, tacking between the ice floes. Without a doubt, here was the *Iakut* and it was sending its sloop to greet us.

The sloop was now close, and we discerned inside it some dark figures in strange clothes. It wasn't a sloop but a Chukchi kayak, and some Chukchi were in it. We evened up with them and they called to us. They drew close and rowed alongside the steamer. There were several men and two women in the boat, all with uncovered heads and in fur tunics over which were slung some large bags made from bladders. Bumps on the women's backs proved to be babies, and the mothers, regardless of the cold and surrounding ice, took them out from time to time to breastfeed them. All their faces were so swarthy and filthy that this was visible from the steamer.

We then passed by the small spit, behind which was little Plover Bay,[2] and headed toward the vessel there. What was this ship? A Russian flag, not a military one but a tricolor, was fluttering on its stern, and its name was hard to make out but then became visible. The ship was called the *Progress*. Strange! A Russian flag on the stern of a foreign name.

We dropped anchor. After several minutes, a steamboat left the *Progress* and came to us. Exiting it was a trio. First was a lively gentleman with an intelligent, nervous face, the engineer Shockley, as he introduced himself, second was Vanderlip, a young, healthy, ruddy-cheeked man, and finally, third was a tall, lean Norwegian—the ship's captain. They first of all asked for the captain and exchanged greetings with him, then Shockley went to Roberts and handed him some documents. Bogdanovich then invited them to his cabin for

---

2. Napkum Spit and Plover Bay are on Providence Bay's eastern shoreline.

tea. It became clear from their talks that this strange ship had also come here to find gold. They said the ship belonged to the Vladivostok merchant Jules Briner, but Shockley—the expedition's leader—said that after reaching Cape Dezhnëv, they'd found no gold along the coast and were now going back. They left after several minutes, but the cutter returned, bringing yet another gentleman in a proper forage cap with a cockade. He proved to be a land allocator who had been taken aboard the *Progress* so as to make certain allocations in Gizhiga.[3] He complained that Shockley forced him aboard the ship and was making him return to Vladivostok. We heard him out, but something was quite odd: here, under a Russian flag, was a foreign vessel out of Vladivostok, as if not knowing that V–ii has already received a concession to the entire Chukotka Peninsula, and here was an expedition under a foreign administration for a Russian merchant familiar to Shockley as well as to Roberts, and here was an official who was forcefully taken to the Bering Sea. Little they said seemed trustworthy.

"In morning, *Progress* gone," in Shockley's words, "to Vladivostok."[4]

That day, we trod upon home soil for the first time. Miagkov and I went ashore with the Americans, but we left them on the beach.

Not only were there no trees, there were no bushes, and underfoot were just pebbles or moss and thin, dirty-green grass with little scentless flowers, and fog overhead. The mountains and the ice-covered sea were lost in the mist. The cold wind blew droplets of water in your face and penetrated the bones. Our surroundings are harsh, bleak, and savage. The sharp cries of geese and gulls and the cracking of the breaking ice are the sounds animating this dead landscape.

A strong gust of wind broke apart the fog cover and there, rising before our eyes, were gigantic black hills bereft of vegetation and covered only by patches of snow. Their summits seemed to be hanging above our heads as if ready to tumble and crush us.

We went toward the hills and began clambering among the shards of sharp stone. Something in the hill deep beneath our feet was grumbling and groan-

---

3. Gizhiga was a district capital on Providence Bay's north shore.

4. Shockley (or possibly Shockly) is speaking broken Russian here. (Further information on him has proved impossible to find.) Akif'ëv's instincts were correct: the *Progress* had no legal right to be there. Washington B. Vanderlip was an American prospector and had previous experience in the Klondike, Korea, and Kamchatka before being hired by the Swiss immigrant and Vladivostok entrepreneur Jules Briner to find gold on Chukotka. Vanderlip, in his coauthored account of his expedition, does not even mention Shockley's name and moreover portrays himself as expedition leader. But he offers sufficient evidence showing Briner sponsored the expedition fully knowing of Vonliarliarskii's mining concession. See Washington B. Vanderlip and Homer B. Hulbert, *In Search of a Siberian Klondike* (New York: Century Co., 1903).

ing. This subterranean noise grew terrifying amid the dead landscape. It was the sound of springs running beneath the rocks.

We'd gotten separated from one another. Suddenly, I heard Miagkov shouting:

"Come here, quick!"

I ran over and saw he was standing and pointing at a human skull between the rocks. We'd ended up in a graveyard. Surrounding us, tossed about here and there, were human bones. The Chukchi greatly memorialize their deceased but inter them very uniquely. After death the deceased remains at home four days, and no outsiders are allowed to enter the yurt where it is. Upon the expiration of this period, the deceased is solemnly laid on a travois and, if it's summer is carried by hand, or if it's winter is dragged by dogs to the hills, where their cemeteries are always located. There the deceased is laid completely naked on rocks and surrounded by an oval of stones. A pipe, small knife, spoon, arrowhead, whetstone, and, if it's a woman, a thimble and comb are placed beside the body as well. The poles from the pallet or travois are laid across its legs. The sledge is broken apart and the deceased's clothing and belt are cut into pieces and covered with rocks. In this way, the corpse is left to decompose in open air. It must be imagined that dogs, usually poorly fed here, do not avoid the deceased, since the bones were strewn all over the hills from top to bottom and there wasn't a single intact skeleton. Based on most of the bones, it could be imagined this cemetery was very ancient and that the village below used to be much larger.

15/28 June. This morning we decided to go on an excursion for gold. I accompanied our crew of gold prospectors. We set off in the cutter, towing the sloop behind us. We ran into a bunch of ice in the bay, driven here by a southern wind. Our small cutter easily maneuvered between the floes glistening in the sun. On the way we scared off an entire flock of birds, which, having circled overhead, immediately settled once more on the water.

We proceeded for an hour and a half toward the mouth of the Olenna River. It empties into the sea at the very beginning of Providence Bay, from its western shore. Reaching the shore was difficult. A powerful surf was pushing ashore the ice field, which was shifting and breaking apart. Its base was littered with rocks. Not finding a calm, peaceful spot, we decided to anchor the cutter and take the sloop itself into shore. But this wasn't easy. Getting into the ice was dangerous; it could surround the sloop so it couldn't get out; the rocks among the surf weren't entirely accommodating and might puncture the hull. We chose a spot where there were fewer rocks, turned the sloop's bow toward shore, and, tossed by the waves, landed on the beach.

FIGURE 4.2. Steam cutter, Providence Bay

The sloop simply grunted and we managed to spring onto land, but not before a wave flying from behind doused us head to toe. The cigars and matches I had in my pocket were completely soaked. Before the others disembarked I grabbed a rifle and went running to the hunt. There were a lot of birds. Whole clouds of ducks and geese were heading toward a large lagoon separated by the sea spit, and they covered it like sandy islands. But they were very cautious and, having noticed me from afar, ascended with shrill cries and flew away. I only managed to kill a few of them.

I ran along the Olenna River's wide valley with much pleasure. The river flows parallel to Providence Bay, southward through the valley, guarded on either side by tall mountain ridges lying in several gauntlets. Going from place to place, our gold prospectors tested the sand and found traces of gold. Everyone was overjoyed. Here it was, our first day searching, and we'd already found it. Our hunger showed it was already late and time to return. Once more we had to contend with the breakers; getting offshore was even harder than landing, and again we were doused by a cold wave, but after several attempts we conquered the breakers and transferred to the cutter. Al'fred Markovich had already boiled a kettle of tea there and we, snacking and chattering, went to the steamer.

I wanted to lie down to sleep but went on deck. I saw Karl Ivanovich standing with a gun, waiting for a seal. I stopped. We waited several minutes. A

FIGURE 4.3.   Olenna River

seal poked its head out of the water. Karl Ivanovich fired, Miagkov ran toward the sound, and we, having grabbed a boathook, went out in our dingy. As it is, only two can fit inside it, and it leaks. We floated around the ship, but the seal didn't appear. This bothered us little, since it was nice to loll on the bay's gentle water amid the white twilight that substitutes for night. And there are many seals here: they poke their heads out of the water here and there and observe the steamer with curiosity. Picard already killed one yesterday.

16/29 June. The weather's been beautiful since morning. The fog lifted, thickened into clouds, and encircled the dark mountains in a light, white belt. On the mountains, a band of snow was gleaming in the sun. White patches of ice abounded on the bay's emerald smooth surface. And above, the sun was cordially spilling its rays from the enormous pale-blue cupola of the sky, as if wishing to compensate Earth for those dull, interminable days when it was hidden in a cold, foggy shroud.

Two Chukchi boats came alongside us. Having learned I was a doctor, some Chukchi were coming to ask for medicine. I examined them and unfortunately found that none were healthy. Most had diseases of the eye, everyone without exception had coughs, and many had a kind of herpes, a scabbing on the skin, covered by a layer of fat and grime. Having met their needs as much as

FIGURE 4.4. Whalebones

possible, I sent them home, saying I would come and examine them further after lunch.

Having lunched, I grabbed the photographic apparatus and went ashore. A scattered heap of whalebones was onshore, proving there are many whales here. I toured all the yurts, examining and photographing the Chukchi. There are all of three yurts and a tent onshore here. It's hard to relate the impression made on me by examining the Chukchi's abodes as well as them.

The yurts are constructed thusly: a spot is leveled in the shape of a circle and ringed by a series of rocks; whalebones are planted in the ground, given the lack of wood for making the yurt's framing. All possible pieces of wood are made do for this purpose. The frame is covered over with walrus or *lakhtach*[5] hides held down tightly with straps. In such way, a circular tent is had, with a peak usually a bit eccentrically disposed. So that strong winds don't blow the covering off, the straps are drawn tightly across the roof and down the sides and secured to the rocks. The yurt's interior is divided into two sections: the sleeping and the general area. The sleeping berth is situated along the side opposite the entryway and amounts to its own bunk, two *vershki* high, a *sazhen* wide, and also as long. The bunk is made from hides and covered in

---

5. Whale.

FIGURE 4.5. Chukchi yurt

several layers, the floor is also covered with deerskins, and a drapery of skins overhangs the entryway. The one bunk serves as a bed for the entire family; if several families live in the yurt, they have several such bunks. The owners' things are placed in the anterior section, sometimes very randomly: furs, meat, and rifles and other hunting gear. The fireplace on which they prepare their meals is here in the middle of the floor, and in this same place they also keep their sled dogs on very nasty winter nights.

Simply approaching a yurt, you're struck by a strong, unpleasant stench. Scattered around the yurt are half-gnawed bones, scraps of animal skins, and various refuse, and you might even encounter a dead dog. Yet, upon entrance to the yurt, this stench becomes worse and more powerful, and you can withstand it with difficulty for only a few minutes. Still, all this is nothing, because you have to summon up an even greater courage to lift the bunk's covering and awaken the head lying there. You're then confronted by smells that immediately nauseate even a strong person. In terms of my strength and wits, the sensation at that moment could compare only to the very height of activity in the anatomy theater's dissection hall.

The berths are illuminated and warmed by seal fat, which is poured into a small cup and burns inextinguishably, releasing insupportable fumes and soot. Chukchi sleep there in a jumble completely naked, men and women together, adults and children, and it's said they sometimes invite their favorite dog to join. Such is the prevailing atmosphere presented you; and you simply need be amazed at how they manage to survive and not suffocate.

Upon entry to one yurt I was immediately shown a certain woman. She crawled from under the cover completely naked. Her head was covered in scabs; her inflamed, suppurating eyelids opened with difficulty; her pale, swollen gums revealed encrusted teeth; and the dry, brittle hair on her head was shaped into something resembling a large piece of felt and looked a fright. Her neck glands were swollen, and in place of a resonant voice her chest emitted a high-pitched wheeze. Her yellow wrinkled skin was covered with a layer of fat and filth. Here was something horrible! She asked me to help her. I promised to do so and pacified her as I knew how, but what could I do there? Our steamer was leaving that evening and entire months would have been needed to cure this woman, because I found she had whole dozens of illnesses.

In another yurt I found a sick woman similar to the first, and also a sick baby of several months; when he turned his face to me I, a man of habit, involuntarily recoiled. The baby's small, wrinkled, red face appeared to be one unbroken scab. And no wonder: he was dressed in filthy furs and his backside beneath his clothing was covered in the same filth; he was still suckling at his mother's filthy breast and didn't know how to wash. And here came the young daughter, and she probably would not have been bad looking but for the suppurating scabs on her ears and cheeks.

Seeing these wretches, knowing I could have helped them, albeit not at all with how they lead their lives, knowing that my knowledge could not be applied, knowing my complete helplessness under the given circumstances, was more than difficult.

I returned to the ship. Along the way I encountered several Chukchi. Conversing with me, they smiled. They can still laugh under such conditions. Living beneath cold winds, amid ice floes, in constant dampness, among bare mountains and rocks, seeing neither bushes nor trees even in summer. Eating seal meat, fat, a dead whale they drag from the sea, sleeping on stinking furs in the filth of a smoky yurt. What for, this life? But honestly, they live, joke, and laugh. Without pretentiousness, this is what can be!

There was much conversation among us on the ship that evening; the Americans were all dissatisfied and said so openly.

Roberts didn't release the Chinese for work today, saying there was no gold here and no worthwhile work and that the Chinese have been hired for manual labor but not groundwork.

There's still no *Iakut*, and Roberts is now saying: "I absolutely don't need the *Iakut*, let it look for us, I'm not planning on waiting for it." Relations have turned sour in general. In the evening, Miagkov returned from the Olenna stream and said that gold was turning up in every panful.[6]

17/30 June. We've stopped at Cape Chaplin within sight of the village of Un-yyn, called Indian Point on the maps. The weather has been beautiful, warm, calm. Having grabbed the Chinese, all the Americans of our company left for shore, boastfully declaring: "This here's a real Cape Nome and where we should look for the gold." After a half hour, we set off behind them.

Walking along the shore, on which there was snow, we passed through the hamlet. The settlement of Indian Point is spread along a flat, narrow spit, its point jutting into the sea; this is the largest of all the settlements we've seen on the Chukotka Peninsula up to now. The population numbers around 300. The yurts here are much bigger and better than on Providence Bay; some of the skin roofs are even covered with canvas. There are even three small wooden buildings knocked together out of thick planks, with framed and glassed windows. These are depots built by the Americans, and inside are furs as well as other Chukchi and American trade goods. Evidently, the Americans visit them quite often. We even saw several dozen large whaleboats onshore and barterers among them. Quite soon an entire horde of Chukchi encircled and escorted us through the hamlet. The Chukchi here are healthier and cleaner than at Providence Bay. Some of the women were rather good-looking; one was not at all afraid and fully able to speak with us quite confidently, fluently, and with feeling. Most of them could understand English, but few could speak it.

The Chukchi are a well-proportioned people of middling height (the men average close to 2 *arshiny*, 5 ½ to 6 *vershki*, the women 2 *arshiny*, 3 to 4 *vershki*); broad-shouldered, strong, and hardy; the skin on their face and body is dark-complexioned; their faces are circular or oval; their craniums are not elongated; and their eyes are dead-set brown or black. Facial hair either doesn't exist at all or, if to be encountered, is very rare. The hair on their heads is straight, black, and very thick. The men cut it in a circle and then clip or shave the top of the head, leaving in this way a corolla of hair a *vershok* wide around the head. The women wear two braids. They all tattoo their faces as follows: on the forehead are two vertical blue stripes extending to the end of the nose ridge, and then, on the chin, two or three upward-curving arched lines or three vertical stripes; such arched lines are also on their cheeks. Besides their

---

6. Throughout the diary, Akif'ëv's descriptions of the presence or absence of gold on Chukotka are difficult to reconcile with each other, suggesting that he had little knowledge of, or interest in, the matter. Bogdanovich, for his part, wrote that because the expedition ended prematurely he could not accurately determine the peninsula's gold reserves. K. I. Bogdanovich, *Ocherki Chukotskago poluostrova* (St. Petersburg: Tipografiia A.S. Suvorin, 1901), xii.

FIGURE 4.6.   Indian Point warehouse

faces, the women tattoo their shoulders and chests. There are no tattooed men.

All of Chukchi clothing consists of hides. The men wear a wide deerskin blouse lined with fur on the inside or the outside, with a wide-open collar trimmed with dog fur and cinched with a belt on which hangs a knife in a leather sheath. They wear narrow trousers of either sealskin with the fur on the outside or deerskin with the fur on the inside, and boots of either deer-leg skins (so-called *kamasy*) or cured and processed sealskin with soles made from bearded seals. The latter are very light and completely waterproof. In winter they wear an additional fur-lined blouse (a *kukhlianka*) and on the head a small *malakhai*.[7] The women's costume consists of the same wide blouse, paired beneath with just wide trousers (resembling a clown costume) and small boots of varicolored hides. On their hands they wear mittens and gloves, sometimes of four fingers. Most of the hemlines of the kukhliankas, boots, mittens, and hats consist of wonderful, ornamentally-stitched designs of small, varicolored pieces of hide that are very pretty and recall embroidery, or they are embroidered with varicolored threads. The women wear beads and love any sort of ornamentation. Seen occasionally are kukhliankas sewn from pro-

---

7. Fur cap with earflaps.

cessed bird skins, predominantly duck, with the feathers on the outside and which are also exceedingly unique and beautiful; or sewn from plucked skins and similar to especially soft, refined kidskin. In damp weather, they pull a wide blouse of bladder skins over their fur clothes, which, though it gets soaked, won't let water through; or, finally, a canvas blouse acquired from the Americans. The children's clothing is the same as the adults'.

In terms of hygiene, such clothing is without doubt unsatisfactory. True, it is warm and appropriate for a severe climate, but since a Chukcha puts the hides directly on his bare body and, once having put them on, wears them until they're practically worn out and, owing to their durability, wears them for a long time and then naturally infuses them with sweat and sebum, they get filthy and cause various skin diseases.

Walking through the hamlet—scattered all over with bones and all sorts of things, even pieces of broken Winchester rifles—we reached the shoreline where the Americans were looking for gold. But they'd found not even an indication of gold, and some of the humiliated ones were slinking back toward the steamer. Along the way, we passed a depot where White had just managed to buy a polar bear fur and several fox hides. In general, our Americans have already proved themselves to be proper shoppers and they managed to buy a pile of furs and other items before we left to see the Chukchi yurts.

We were dawdling through the hamlet. Suddenly, I heard a long, drawn-out whistle and then shouting: "Doctor, to the ship, there's been an accident!"

I ran to the shore, where all of us were now hurrying to from every direction. Everyone was jumping every which way into the sloops and making for the steamer. The flag was at half-mast.[8] Everyone was anxiously wondering what had happened.

It turned out to be very simple. Roberts got tired of sitting around and decided to raise anchor; therefore, he gave the order to blow the steam whistle to call everyone to the steamer. Karl Ivanovich lost his temper. Indeed, the impudence was enormous, for he'd issued his order knowing nothing of the vessel and without informing the expedition leader. In general, our Americans have little self-restraint.

Karl Ivanovich delivered a reprimand, ordered the steam shut off, and, in exchange for this, allowed a pair of cutters to go investigate the coast north of Indian Point. I didn't go with them; but having grabbed the photographic apparatus, I went again to the settlement.

The Chukchi were intrigued by the photographic apparatus and once more encircled me in a dense crowd. I can't say this was pleasant, since the offensive

---

8. A flag at half-mast is a nautical sign of distress.

stench of cooked fat emanated from them, but, like children, they were so profoundly overjoyed at seeing the miniaturized form of an object in the frosted glass, and they posed with such readiness, that I spent a long time with them.

Surrounded by a group of boys, I went and sat down to relax on the shore dotted by enormous chunks of ice. All along the shore Chukchi were sitting with rifles in their hands a certain distance from one another, but at the swash itself they were getting ready to cast their light boats into the water. This hunt intrigued me very much.

The Chukchi are superlative marksmen. As soon as a seal's head appeared above the water several dozen *sazheni* away, a Chukcha would raise his rifle; the wounded seal would spin convulsively, blood spraying in all directions. It dove, but a light dugout would immediately fly like a bird from shore and, standing on its prow, a Chukcha with a harpoon in his hands would be looking keenly from side to side. The wounded seal would appear again above the surface and be struck by a bullet and then harpooned by the Chukcha. The hook would pierce deep into the creature's body and a spear thrust would end its life. The hunters' speed and dexterity had to be seen, and sometimes they were really just youths of 12 to 15.

Returning to the steamer, I called upon a certain Chukcha who knew several words in English, and he showed me his weaponry. They are now armed with the .44-caliber rifled Winchester, which they get in great quantity from American whalers in exchange for furs. In every yurt you'll find a Winchester, sometimes several. For hunting seal and walrus they use a harpoon in addition to a rifle. This they make very cleverly. A wooden pole, 2–3 *arshiny* long, is tipped with a small cone of walrus tusk that also has a bone end-piece with an iron point, from the end of which there runs along the pole a long, slender strap of sealskin, whose upper end is connected by a loop; when the harpoon is cast, the tip pierces the body and stays there, and the pole falls aside and floats. In this way, the seal is stuck on the hook and restrained by the long strap. For the final blow, Chukchi use either a lance with an iron tip or arrows. They have bows and arrows as well, but the rifles have replaced them almost entirely. For hunting birds when they fly overhead in a flock, the Chukchi use a special missile tool. It consists of five rather circular pieces of ivory about the size of a walnut attached to a string made of veins, the ends of which are connected. They throw this gadget straight at the flock, and after expanding in midair, it ensnares or hits a bird, which the Chukchi pick up and finish off.

During suppertime in the evening, the steamer was besieged by Chukchi coming to proffer their goods. A group of women huddled in the dining room's doorway. They grabbed the leftovers from our plates and greedily devoured them. They took everything that remained of the meal, laid it straight on deck, and with unfeigned pleasure quickly stuffed the pieces of food into their mouths using their hands.

Our miners returned in the evening; they'd gone nearly as a far as Cape Mertens. The response from them was disappointing: though the strata along the coast proved completely similar to the strata at Cape Nome, they found no gold.

18 June / 1 July. There was a drama this morning. Roberts, extremely dissatisfied with Karl Ivanovich, began quarreling with him on deck. They quarreled quite loudly and energetically until Karl Ivanovich informed him that if he didn't stop, or if he repeated this, he would have to be dropped off. Discussions ensued and Roberts and Reiner even demanded their passports, saying they didn't want to be in the expedition and couldn't go on with us. But they were persuaded this was impossible and have apparently reluctantly resigned themselves.

By lunchtime we were already in the Seniavin Strait and entering Glazenap Bay[9]; going further into the strait was impossible, it was full of ice. Therefore, we hurriedly inspected the bay's coastline. Wandering along shore, we noticed a solitary cross standing on a bare, rocky hillock not far from the waterline. An iron plate with a brief inscription was nailed to the cross: "Here rests Egor Purin, clerk of the clipper *Gaidamak*. Died 30 July 1875."[10] It was quiet, deserted, deathly. Mountains, bare and mute, encircled and guarded this solitary grave, and the cold, transparent sea covered in ice lapped at our feet. Yes, he was at rest here; here was a peaceful death. And hiding between the rocks, just near the cross itself, was a small bundle of pale forget-me-nots, as if voicing that he who lay beneath this cross, despite being anchored here many years already, was recalling his far-off homeland. This was something terrible and sad. Perhaps one of us might end up staying here.

Leaving the bay, we turned back and headed north, passing east of Arak Island.[11] More ice everywhere along the coast and filling the strait of Mechigmen Bay necessitated we go further. By morning, we were off Cape Novosil'tsev at the entrance to St. Lawrence Bay.[12] But this bay was packed with ice, so we dropped anchor several versts from the cape.

---

9. Seniavin Strait, named after Admiral Dmitrii N. Seniavin (1763–1831), is on the Chukotka Peninsula's southeast coast. There were several prominent Glazenaps.

10. The *Gaidamak* entered service in 1860 and was part of the Siberian Fleet.

11. "Arak" is an abbreviation of "Arakamchechen."

12. Cape Novosil'tsev was probably named after the statesman Nikolai N. Novosil'stev (1761–1838). St. Lawrence Bay was named by Captain James Cook, who arrived there on the Feast Day of St. Lawrence, 1778.

FIGURE 4.7.   Cape Novosil'tsev

Our American companions have been eyeing us like wolves. I was standing and eating with Miagkov at the buffet this evening. White and Landfield came over and we began a conversation that soon moved to the topic of the day, and we said many unpleasant words to each other. It's become clear we differ completely in our understanding of the expedition's goal and planning, that the Americans consider Karl Ivanovich just a figurehead and recognize Roberts as their leader and themselves to be in charge of the expedition. We've come to fully understand the matter is not going to go well for us and that we Russians can expect nothing from the Americans. They've become disgusting and at the same time so absurd.

You should see what a huge role ghosts have been playing in our expedition! Mysticism lies at the heart of our expedition. A year ago, our carpenter Charles Scranton turned to a fortune-teller for some reason. Through her the spirits informed him he would go on a profitable job far away, but he would not get rich and would have to leave, because one of the people closest to him would die. But he would have occasion the following year to go even further, to an unknown region, and there he would find wealth. He remembered this prediction. He went to Alaska and was working there last spring; suddenly, he got a message that his mother was ill. He abandoned work and went to New York. His mother was dying. After some time he heard around that a lot of gold was being found on the Siberian coast. By chance he bumped into Roberts and, having seen half the prediction already come to pass, decided that Siberia was the land he'd been promised to find wealth in. He's shared with Roberts what the spirits foretold.[13]

Roberts, an insane old man, sees a good omen in this, and Scranton is becoming our expedition's Mascotte.[14] Such faith does he have in his own lucky star that he allows no one to step foot onshore before himself. This enormous figure, with shovel in hand and pan under armpit, debarks awkwardly, and immediately begins digging and tossing sand. Of course, he doesn't manage to find gold, and categorically declares: "No gold here," then walks just as importantly to the sloop, the rest of the Americans trailing behind him.

I now understand why one of the Russians in San Francisco called our expedition a motley crew. Indeed, who is not among us: Russians, Englishmen, Americans, Jews, Norwegians, Chinese, Poles, Mexicans, Finns, and even one Ossetian.[15] In addition, it's similarly motley from the business perspective. There are two engineers: Karl Ivanovich and Rickard, very well, but as a former mining student, Aleksandr Gennad'evich is also doing this job. But who else? Roberts is a crippled old man whose eyes close such that we've dubbed him Vii: "Lift my eyelids, I can't see."[16] It's said he's an expert in the matter of gold, but he apparently knows it on the exchange and not in the sands. White, a former diplomat, is quite in his cups now and has appointed himself a master of capital. Stern is also a capitalist. Dowlen has never been involved in the gold business. Scranton, the father, is a carpenter, and his son worked in a cannery sealing jars. Picard is a stuffer of animals[17]; Landfield—you can't tell, since he says he's earned two degrees, served in the army up to the rank of captain, and says much more about himself. But his appearance doesn't correspond to his titles. I've noticed another curious thing. Our Mister Stern titles himself a major. Turns out, he was never in the military but *is* a major in the Masons. We have more. Our captain, Jahnsen, is a Mason of the 32nd

---

13. During this time, belief in the paranormal and supernatural was common throughout America and Europe, and Russia as well.

14. In Russian, маскот (*maskot*) translates as "mascot." However, Akif'ëv writes here "Маскота" (*Maskota*). This word's capitalization and vowelized (feminine case) ending suggest a reference to the French composer Edmond Audran's 1880 opera *La mascotte*, for which Audran idiosyncratically used the term to describe a farm girl whose mystic powers brought good fortune to those around her.

15. Akif'ëv nowhere specifies who in the expedition was Mexican or Finnish, so he may be exaggerating a bit for comic effect.

16. A famous line from Nikolai Gogol's story "Vii" (1835), in which he portrays as King of the Gnomes a character from Slavic mythology who judges and punishes people for their misdeeds. At one point in the story, Vii orders visitors to lift his enormous eyelids off the ground.

17. I.e., a taxidermist. Akif'ëv himself will soon reveal he is a taxidermist, so he is obviously belittling Picard's skills.

degree, and he's terrifically proud of this, always showing off his cross that hangs on his bracelet, and proudly says he's achieved a high Masonic rank and that higher than he is only a certain Prince Welsky, who is a Mason of the 33rd degree. He showed us a photo where he's pictured with his young clerk, also a Masonic general. Jahnsen often boasts about his own physique.

There are still more Masons among us expeditionaries, only of lower rank. Some sort of dangling medallion hangs from their pocket-watch chains. While still in San Francisco I heard about American Masons from Fedorov. He said this society is widespread in America and is sufficiently tightknit; but it's clear from his stories that these societies don't at all resemble the former Masonic lodges of Europe. There's nothing religious to them, it's simply a mutual aid society to which members contribute certain sums for which they receive elevation through the ranks. Otherwise, it's hard to see how such an utterly stupid and vulgar man, and such a drunkard, as our Jahnsen is could receive a higher degree.

19 June/2 July. We're still off Cape Novosil'tsev.

Karl Ivanovich went to shore with Miagkov and Al'fred Markovich in the sloop, but I busied myself trawling. I have a big problem: I can't find where the fishing tackle I ordered in San Francisco has been put. It was paid for and should be somewhere in the hold, but my searches have been in vain, so I've been making use of an iron ring covered in netting. It's quite awkward, though I caught several crabs, shrimp, and one hermit crab today. I didn't see any fish at all. They returned after three hours from the coast and told us something interesting. They discovered a whole party of laborers onshore who were transferred from the *Progress* and are living on the beach in a tent. It transpires that Shockley duped us, as did the Russian administrator who was with him: they said they had finished prospecting and were going south. In fact, it turns out they transferred the party to go prospecting and, two nights ago, came here to fetch them, but the ice was so thick along shore the *Progress* couldn't put in and turned back. This party has been searching for gold here since the 21st of June. It's been generously supplied with coal and provisions. They obviously basically counted on working here.

An hour later, two from this party were with us aboard the *Samoa*; what was my surprise when I realized they were Bakharev and Malafeev, travel companions in our expedition to northern Korea that we completed in autumn 1898. Who could have said we would meet again here, just below the Arctic Circle? Compared to this pair of strapping young men our Chinese were simply pygmies and sad to look at. Even the Americans were similarly diminished.

We invited them for a celebratory beer and began talking. It turns out that Shockley managed to hire workers much better than we did. In Vladivostok he hired Russians at 30 rubles a month, and Chinese and Koreans for much less than that, whereas we're paying up to 60 dollars a month to weakling Chinese unsuitable for anywhere, and especially unsuitable for our circumstances. To the question of whether they've found gold, Malafeev answered that till now they have not, but they did find a lot of black and red sediment.[18]

Saying good-bye, I supplied them with Cinchona, of which they were low on, and they in turn sent me a boat of Bogdanov tobacco, which very much delighted me since I don't like the American tobacco.[19]

In the evening Miagkov and I, after everything aboard the steamer had settled down, brought a bottle of California wine, some raisins, and canned pineapple to the bridge and enjoyed ourselves. The sky was an ashy blue color girded by a pink ribbon of sunset, and the fresh clear air, the white ice floes' fantastical shapes on the sea's steely-smooth surface, the dark, grim mountains off in the distance, this soft, white world—all this was so dissimilar to night that we did not want to sleep at all, and the floes and the pineapple with the vintage wine put us in a cheerful mood, in contrast to those mountains. We proposed toasts and dispatched heartfelt greetings from the Arctic Circle to our dearest ones still in distant Europe.

I wanted to test the sun's power, and at 12:00 o'clock sharp took a photograph of one of the Americans creeping across deck. It was a 3-second exposure and the print came out strikingly clear.

20 June / 3 July. Having breakfasted early, we went ashore in the sloops. The ice was still nearly packing the beach; we tacked between the floes and sometimes encountered dead ends, making it necessary to turn around. Our gunwales just barely eluded the tall floes that melt from the bottom. Sometimes such floes would bulkily flip over and plop into the water in front of our bow. Luckily, we got between them and landed on the beach.

I headed for Malafeev's campsite. He's constructed it wisely: not far from the village of Gidonei, he's staked a tent to a dry, flat ledge along the shore with the bordered side facing the sea, inside he's spread out a plank floor and set up his bed, and in the front area he's placed a stove on a slab. To top it all off, he's erected a Russian tricolor flag above the tent.

---

18. Black and red sediment possibly indicates gold.

19. The bark of the Cinchona tree is a source of quinine and is used to reduce fever. In 1900 the St. Petersburg tobacco firm Shapshal Brothers went into business with A. N. Bogdanov and Co. to create the Laferm ("The Farm") tobacco factory, which became Russia's leading supplier of tobacco and tobacco accessories.

Having jabbered with him, I went to the mountains, where, I'd been informed, two recently deceased persons were lying. I found them between ledges on a mountain summit. They had already begun decomposing and the smell was quite sharp, so taking a photograph wasn't especially pleasant. Overpowered by disgust, I took the photograph and was even guilty of the little crime of robbing their graves: I took those objects the Chukchi place around the deceased. I saw they were each lying alike, the head pointing north. My eye was caught by a very unpleasant circumstance. Water flowing from the cemetery was running into the brook that goes to the village of Endogai[20] and from where it gets its water: it can't be said it would be good to drink the water there.

I returned from the cemetery to the campsite once more and they invited me to lunch. We drank diluted spirit by the glass and enjoyed cabbage soup with hen. It was canned, of course, but I liked it very much since I've gotten quite sick and tired of the constant soups of pickled oysters or condensed milk. As it were, we formed a quartet of those who were previously on the Korea expedition, and somehow we immediately began recalling our encounters and the journey's difficulties, though more so its numerous curiosities, and we were transported again—away from these ice floes, the tundra, the smelly yurts, and the beast-like Chukchi—to where rice and plague-ridden paddies flourish, to where emerald mountain rivers babble among proud cliffs, their spiney summits rising high into the blue sky to be embraced by delicate white clouds. We remembered the rich forests, the simple, clean fanzas,[21] the kindhearted Koreans in their broad white costumes, and were brought there once more.

However, back to those Americans. Yesterday they consulted something and now here they are very obligingly stopping and suddenly changing tactics; but this doesn't comfort me: *quidquid id est, timeo Danaos et dona ferentes.*[22]

I was busy sorting out Malafeev's apothecary after lunch when some Chukchi arrived. Some of them had been treated very successfully by the workers. Here is one such case. There came a boy whose entire cheek was a kind of herpes, and the workers first of all washed him with soap, because he was incomparably filthy, then cleaned the herpes with carbolic and applied boracic vaseline to it. Two days later, the herpes had almost completely disappeared. Seeing this, the Chukchi began washing their faces and pleading for soap. They'll obviously imitate many good practices they see but there are no exemplars. Russians almost never visit at all, once every several years, and the

---

20. The name Endogai shares the same letters with the previously identified village of Gidonei, so this may be another example of poor copyediting or sloppiness on Akif'ëv's part.

21. Fanzas are Korean peasant huts.

22. "I fear Greeks bearing gifts, whatever they are." Virgil, *The Aeneid* II: 49.

Americans only come to trade. One of the Chukchi who came, and who considers himself the village headman, complained, quite thoroughly so, that the Russians have forgotten them.

"We know there is the same tsar for us Chukchi and for you Russians," he said. "But you don't visit us, and the Americans bring us flour, guns, and tobacco. What would we do without guns? We know the Americans better and understand their language, but Russian, we absolutely don't know."

And he's absolutely right.

Contrarily, there should be, similar to the American States, trading cruisers that arrive on our eastern shores annually. Many of us in Russia have forgotten our godforsaken territories, yet the Chukotka Peninsula is evidently the most wretched of them.

One little Chukcha patient is absolutely at ease around our workmen-physicians, and has even begun learning how to play checkers and to play it sufficiently well. I enjoyed playing a game with him.

Al'fred Markovich arrived, and we went under sail in the sloop to the cape's south side, where Karl Ivanovich and Miagkov were awaiting us. They'd been unable to get through the mountains and were terribly worn out. Having grabbed them, we sailed smoothly through the ice to our *Samoa*.

After supper the Americans inveigled to fire their revolvers at some bottles. They were indeed far worse than the Tarasconians.[23] They fired around a hundred cartridges but were unlucky. Exhausted by the day's hike through the mountains I, regardless of the shooting outside my cabin window, drifted into the deepest sleep.

22 June / 5 July. We're going to Cape Nome. The weather is deteriorating again. The wind's blowing, it's cold and foggy.

Yesterday, having given Malafeev an order to cease work, which he heeded as being rather expected, saying that Shockley had warned of this, we raised anchor and passed through Seniavin Strait's north end. The day was clear and warm, 12 ½° R in the shade. The sea was smooth as a mirror, though there was still a lot of ice in the bay.

It was decided to disembark Miagkov, with three Russians, the Finn, and a couple of Chinese, at the mouth of the Marich River,[24] and go by ourselves to Cape Nome, since we still haven't found the *Iakut* and the Americans steadfastly continue demanding they be taken to Nome immediately.

---

23. Residents' poor hunting and shooting skills feature in the French author Alphonse Daudet's satirical novel *Tartarin de Tarason* (1872), set in a real-life Provençal town.

24. The Marich River is at the north end of Seniavin Strait.

The shore at the mouth of the Marich is low and gently sloping, behind it is a moderate rise and a covering of tundra, and farther on, mountains. There was still a lot of snow nearby, but none on the grass where a swathe of various flowers with extremely pleasing scents was scattered. We found fragrant forget-me-nots, distinct from our short-stemmed ones and much more intense in color. In a small cove on the same beach we staked out two tents, laid down plank floors, and hoisted a Russian flag above the tents. Thus well done, it became comfortable there. I didn't want to return to my cramped cabin or see our fellow passengers, so I stayed there till morning. We made a campfire, heated up the teapot, and, placing it in the sand, sucked the warm tea down with brandy. Three Chukchi yurts were near our tents, and a Chukcha approached us. He was only able to communicate using sign language. When we offered him some tea, he sipped from the tankard enthusiastically, but did so wincing terribly from pain. All his teeth were chipped, and the flesh on his tongue and gums was missing. The poor man was obviously having difficulty with sustenance. We diluted some potassium manganese in a bottle and ordered him to gargle it, and provisioned him with tea, sugar, and bread. Having drunk his tea, he nodded his head in a token of gratitude and dragged himself home. I, too, bid my farewells and left for the *Samoa*.

There was a drama last night. Like every night, the captain got drunk and fell fast asleep. Just before morning, Lemashevskii, having woken, went on deck and took a fright. The north wind was pushing us through the strait, directly into a huge ice field. It was already close and had to be averted as soon as possible; Lemashevskii woke the captain and told him of the threatening situation. He went on deck, looked about with bleary eyes, muttered that he'd commanded a ship for 17 years and there was no danger, and returned to his cabin. But all the ice shifted; several minutes later, it was crushing the anchor chain. We ran to awaken the assistant captain. He then immediately shouted an order to raise anchor. Another two minutes, and it would have been bad. In the best case, we would have lost the anchor. Ice was now blocking the strait's northern exit, so we turned south and headed toward Indian Point. There, we were told there'd been no sight of a navy ship. That meant the *Iakut* still hadn't arrived. It was left to give in to the Americans and leave for Nome.

We've taken aboard two Chukchi from Indian Point with trade goods they're bringing to sell in Nome.

CHAPTER 5

Alaska

Nome—The gold business—Memories of the sale of the North American holdings

23 June/6 July. Nome. Going on deck this morning I saw something unusual. Stretching in a line for many miles along shore was a real encampment. Tents, tents, tents, how many thousands of these tents it was impossible to say. There was an entire forest of masts of various steamers, barges, and schooners filling the roads. This is Nome, the golden city that has sprung up over several months. The shoreline is gently sloping, sandy, with no cove at all, and the vessels sit in the open sea.

We dropped anchor two versts from shore and went to the city in a sloop. Crossing the breakers, we entered the mouth of the small Snake River and reached the wharf. The animation was terrific. People were knocking into me, and the street was so crowded I could barely pass! Buildings covered with signboards and sometimes three stories tall stood like houses of cards along both sides. Everywhere you looked were American flags. The noise, talking, and crowding in general made it like a holiday fete. You couldn't force your way down the wooden sidewalk, and walking in the dirty road, you had to beware that a cart beam didn't hit you in the head. Horse-riders and dog-walkers were practically nonstop. It seemed odd that such a throng of people was meandering down the street, smoking, hands in pockets, migrating from saloon to

FIGURE 5.1. Nome beach tents

saloon. It might have been expected that gold fever would make everyone greedy to work, but that wasn't evident here.[1]

Through filthy mud puddles we reached the Alaska Commercial Company store, asked if we couldn't write home, and then returned to see the city and the works. But of the latter, we saw very little. Several fellows were digging in a spot on the beach, and rinsing with pans or small rocker boxes. The gold-bearing sand was close, all of two feet from the crust, and this so-called red sand, or *ruby sand*, consists of small pieces of granite. Gold, though not especially abundant, was turning up in every sluice: you were simply amazed that so many people weren't working. Pits have been dug everywhere on the beach; though it's now become quite impoverished since the earlier years, because the work is done with harried hand, it's of course quite possible that much gold is still here. Several rows of tents stretch haphazardly along the entire shore. There are also signboards for various shops, pharmacies, photographers, canteens, doctors, lawyers, and whatnot! They're all well designed. Especially those for the lawyers and doctors. There's lots of business here, es-

---

1. Akif'ëv seems to have been oblivious to the fact that they arrived two days after Independence Day. Nevertheless, Nome would have been a very crowded place at this time.

FIGURE 5.2.   Nome prospectors sifting sand with a rocker

pecially for the latter. It's no surprise that people living in a tent beneath cold winds in the constant damp get sick here.

We reached the quarantine, surrounded by a shallow ditch and denoted by yellow flags. A smallpox epidemic has been spreading widely here. It's said that many are dying from pneumonia and forms of typhoid.[2]

Along the way, we dropped in for a bite at one of the restaurants, which was nothing other than a regular tent, albeit of large proportions. Our breakfast was prepared on a kerosene stove; the entire breakfast consisted of a couple eggs and a cup of coffee with some bread, and for this we paid 50 cents—i.e., a ruble. From there we crossed over the Snake River bridge, paying 10 cents to cross, and came to a railroad that has been built here already. The railroad stretches 5 miles along Anvil Creek and the train takes 30 minutes—a passenger pays 1 dollar and a small amount of baggage costs 2 cents per pound. You look outside and are astounded. All of six months have passed since gold was discovered here, but there's now a city of 20 thousand residents, not to mention the suburbs, railroad, and several large firms' electric lighting, and now

---

2. Smallpox and other transmissible diseases swept through Alaska during this time, though the victims were overwhelmingly Alaskan natives.

they're building a waterpipe from Anvil Creek and, it's said, will next year lay a telegraph line to the nearest city. This is so American.

Having gone back to the main street, which we jokingly called Nevskii Prospect,[3] we encountered by a saloon door our Captain Jahnsen, in an already quite inebriated state talking dazedly about something. All the crush and noise had already gotten on our nerves, and we wanted to return to ship. But some unpleasant news awaited us there. The water had receded, and the river's mouth had gotten so small the sloop couldn't pass through. We had to sit and wait until evening for the tide. Having sat and waited awhile on the street, while our flanks suffered a haphazard massage, we repaired to a café to eat.

We found the Café Royal in a large wooden barn, satisfactorily clean, and appointed with little tables with clean tablecloths. The prices here were intolerable: it cost 1 d[ollar] 50 cts—i.e., 3 rubles—for a steak, and everything was like that. We ate and wondered what to do next. Bogdanovich and his wife decided to stay onshore for the night, but I was now fairly tired and bored with shuffling down the dirty street.

Vasilii and I went to our sloop and, stretching out in the bottom, quickly fell asleep. Whether we slept a long time I don't know, but we were awoken when there was some sort of crashing inside the sloop. The captain turned out to have drunkenly tumbled into the sloop and, unable to get his balance, had crashed through a bench. "What next for our oafish captain?" I thought. I watched him grab the rudder, stand in his spot, and order the sailors who had accompanied him to man their places. The water was now rising a bit and he, regardless of a terrible wind off the sea, decided now was the time to sail the boat. We set out, and luckily crossed the mouth.

But then the drama began.

A bar has formed in the river's mouth, and we had to hug the shore to get around this. The wind was whipping gigantic waves from out of the sea and near shore, the breakers were roaring. The sloop's gunwale ended up facing the waves, and as the first wave was pushing us toward shore and we struck bottom, the rudder came loose, leaving just the tiller in the captain's hands. We almost turned over right there, and the captain was cursing and talking so disjointedly that the sailors began taking matters into their own hands. Straining every nerve, they somehow managed to right the sloop in the cut of the wave; we then got through the breakers and headed out to sea. Dense and curling with gray crests, the breakers were left behind and we pitched into the sea's large, heavy waves. It was now possible to calculate what we were making for, but another incident occurred. The wind blew the captain's hat off and the water carried it

---

3. Nevskii Prospect is St. Petersburg's most famous thoroughfare.

toward shore. The captain simply lost himself: "Goddam, back to shore." Though not especially wanting to go back, the sailors obeyed, set to rowing, and soon pulled us in to shore. There was the hat. The captain tried to snag it with a gaff but, poking the gaff in the water, couldn't even touch it, and we were carried all the way to shore, where a crowd of people had gathered to watch our maneuvers.

The hat was snared, of course, but now we were among the breakers. Shouting "Forward!" the captain cursed once more and grabbed an oar, but the sloop, instead of going forward, was being hit on its side by the waves and shifting awkwardly. In an instant, a wave flew over our sloop, slapping our backs and knocking us into the sandbank; the next wave turned the sloop keel up and we all flew into the cold water, the wave catching us up and flinging us onshore, where five pairs of hands grabbed us and the sloop. We were now out of danger, thank God, but had gotten soaked to the bone as a result. Still, it was good there were no rocks on the beach and the shoreline was close by, for had we been farther from shore nothing could be said of our getting back. The cold was terrible, simply teeth-chattering, and we flung ourselves at a run into a tavern to drink glasses of brandy.

It's relevant to add that taverns here stay open all night and are always full of people, and there are two shifts of servers, each lasting 12 hours. Buffet attendants earn 15 dollars a day—i.e., 450 dollars a month. That's a salary no one could refuse.

Nome's main street comes alive at night. Everywhere there's the sound of singing, music, and gramophones. There's even a theater here, albeit also characteristically American. In the taverns, underneath the hoarse singing of sufficiently lubricated vocalists whose faces are sometimes adorned with bruises, the public plays cards, dice, and roulette. Dollars pour by the fistful from one pocket to the next. There are scales behind the counter for weighing the gold dust that prospectors immediately fork over to the tavern keeper. The priciness, of course, is terrible. "All drinks 25 c[ents]" is written on the sign—that is, ask for a glass of whiskey or a small mug of beer, and you hand over the same fifty kopeks.

Having knocked about and warmed ourselves up in the tavern, we had a bite at the Café Royal and went to our room to sleep. Al'fred Markovich had just managed a bit earlier to get the last available room in Nome, and it turned out to be on the third floor, or, to put this more accurately, it turned out to be simply the attic beneath the roof. The room had no ceiling and the roof declined very steeply. The walls were made of narrow wooden slats. In total, the furniture consisted of a single extendable bunk, a stool, and a washbasin. All this comfort cost us 3 dollars for the night, even though Al'fred Markovich and

I had to share the bunk. There was no stove, of course, and I still couldn't get warm for a long time after the accidental sea bath.

24 June/7 July. In the morning, having drunk coffee, we took ourselves to Anvil Creek. We paid 30 dollars for a two-horse carriage there and back. Such are the prices here for a carriage. It's not too far from Nome to Anvil Creek, 7 or 8 versts, and the road passes through tundra. The wet loam, or peat, kept giving beneath the carriage and the horses were getting stuck and walking step-by-step. Had they not been such good horses they couldn't have pulled us out of there. Indeed, poor ones should not be brought here: the transportation of a horse from San Francisco to Nome is terrifically expensive. Our cabbie explained this issue to us as follows: "I paid 1,000 dollars to transport a pair of horses, a wagon, and a ton of oats, but it was profitable. A pair of horses can earn up to 300 dollars a day, you know. A pair of horses costs 10 dollars an hour, and a dog team is 5 dollars; but now, a lot of horses have been brought up and the pay's dropped to 30 dollars a day."

We were at the mines in Anvil within two hours. The mines trace a creek that flows through a narrow valley and empties into the Snake River. Limestone, micaceous schist, quartz, and a small amount of gneiss and granite prevail among the various strata to be encountered in this stream's valley.

The deposit of gold is huge. Gold-bearing sand is on the surface, but the spoils come from the depths. In front of us, a gold prospector scooped up one trough of sand as an example. There proved to be nearly a dollar's worth of gold and the nuggets were quite large. You could imagine how much would be gotten by rinsing away that body of sand. The gold here is rinsed in a sluice. A bit above the mine, the water for rinsing runs down a thick canvas sleeve and then passes into a long chute, on the outside of which transverse boards are connected. Workers shovel the gold-bearing sand into the chute and the flowing water hits the sand, knocks aside the pebbles, and the gold gathers in the transoms. Gold is to be found in the substrate (the bedrock) and in the stratum of red and black sand. A great amount of very weighty magnetic ironstone is encountered. In the thick parts of the sand the gold deposit reaches a distribution of 3 *zolotniki* per pood. This is a rather insane figure! According to a report by an agent of the Washington mint, Anvil Creek provided E. Lindblom, one of the Anvil Creek mine owners, some 150,000 dollars in 1900, and it reports that no less than 1,800,000 dollars was mined from this stream.[4]

---

4. Akif'ëv cites Bogdanovich as his source for this information: K. I. Bogdanovich, *Ocherki Nome* (St. Petersburg: Tip. A. S. Suvorina, 1901). The Swedish American Erik Lindblom (I have corrected Akif'ëv's misspelling of his name as "Lindbolm") teamed up with two fellow Swedes to establish the Nome Mining District.

Workers in the mines receive 5 dollars a day, from which 2 dollars is deducted for maintenance. Actually, maintenance here is fabulously expensive.

Returning from seeing the mines, we found a small, dirty tent that had been dubbed a restaurant, though it was set in a bog and in no way resembled a restaurant: it had no floor, and the filth was horrifying. We paid a dollar for a breakfast consisting of a couple eggs and a thin slice of ham. Given such prices, how much would a proper three-course meal cost?

During the return trip to Nome through the tundra, we also saw in the tundra test shafts and posts on which it was noted these parts of the tundra belonged to the Tundra Mining Co. The surveying of the tundra has only just begun. Gold has proven to be here, and in considerable quantities. At a depth of 15 feet, the gold deposit varies from 2 to 10 *zolotniki* per 100 poods.

We reached the shoreline quite early during an ebbtide, so once again we ended up wandering through the city. We found a place to mail letters. The office was situated in a small, filthy barracks: it was packed with a crowd of people and another crowd was even standing outside. Everyone was sending money home. To calculate the amount of money coming out of Nome, it suffices to consider a single figure: on 3 July, 68 money transfers totaling 64,108 dollars were completed.[5]

After the post office, we found a photography studio (there are several here, and they do a very good business); we bought ourselves a photo of Nome, and of course paid dearly; along the way we acquired several ounces of gold from one of the shops. It costs 16 dollars an ounce here.

Having eaten a square meal in a restaurant for 1 dollar 50 cts., we returned to the ship in the evening. Unpleasantness awaited me there. Some birds I'd killed for my collection and left tied to the bridge's yardarm had disappeared. The Americans blamed the deed on Roberts's puppy Spot, from which he's never separated and won't even allow on deck without a blanket. But something suspicious is going on. My birds have been disappearing quite often.

25 June/8 July. We're now off Nome; the weather is cold, windy, and damp.

This evening, we decided to leave the boat. The Americans went ashore with the Chinese, 6 miles from Nome, to get acquainted with the practice of gold-rinsing here. Roberts, White, Dowlen, and Scranton stayed in Nome to further study the distribution of gold deposits, but the rest of us went walking along the beach. After having seen everything, it became tiresome.

---

5. During this past July and August [1900], 262,000 dollars' worth of money transfers, or an average of 11,000 dollars per day, was sent—Bogdanovich. [*Ocherki Nome*; Akif'ëv's note; I have corrected his misspelled title.]

Actually, these shores were ours at one time: Russia owned Alaska, the Aleutian Islands, and much else, and in 1866, all this was sold to the United States for 7,200,000 dollars.[6] Russia was merely left with a narrow strip of sea. Only now is it clear how unfortunate that sale was. Before the sale, all this enormous expanse was regarded as a "fur district" and it generated an insignificant revenue. But beginning in the late 70s, there developed a fishing industry, and now, by a conservative estimate, the fish taken from here equal close to 5 million dollars a year. Following the depletion of fur seals the fur industry collapsed, yet by itself, the Alaska Commercial Company enterprise paid the United States government a onetime lease payment of 7,000,000 dollars for the period 1870 through 1890.

These numbers mean nothing in comparison to the figures concerning annual gold production. Let's take the Klondike. In 1899, 16 million dollars' worth of gold was produced in the Klondike, and the amount is even larger for this current year. Over the past year, in the Nome district alone, 2 million dollars of gold was produced; mines are now sprouting like mushrooms; and gold prospectors have gone north and already discovered gold at Cape York and on Clarence Bay. And if the sum of revenue annually received from these former Russian holdings is taken into account, then it reaches a round figure in the several tens of millions. The energetic Americans are now filling those previously nearly empty territories populated by Eskimos, Aleuts, and other savage tribes,[7] new cities are growing, and life is hitting its stride. It's tearfully painful this whole region has been irreparably lost to us and that, really, long ago the existence of gold was actually known about. During the early 'fifties the mining engineer Doroshin, on commission from the Russian-American Company, discovered gold on the islands of Sitka and Kodiak and on the Kenai Peninsula. Having found signs of gold, he persistently recommended carrying out further investigations in the complete hope of good results, but the matter petered out.[8] And now we have such places as, for example, Kamchatka: gold has been found, but it's not being exploited and the land is still wild and unsettled.

---

6. Akif'ëv is correct regarding the cost, but he misstates the date. The Alaska Purchase Treaty was formally concluded on 30 March 1867 (n.s.).

7. The phrase used here, "и те почти пустынные прежде края, населенные эскимосами . . ." is instructive, as it indicates Akif'ëv's understanding (shared among many Europeans and Americans at the time) that the presence of native peoples did not contradict a view of their lands as being "empty."

8. Pëtr P. Doroshin (1823–75) led prospecting expeditions to Alaska's Kenai Peninsula from 1848 to 1851, but he failed to find commercially viable amounts of gold there. Doroshin believed more gold could be found further up the peninsula, but the Russian-American Company's lease agreement with the Hudson Bay Company prevented his exploring those areas. Doroshin subsequently turned his attention to prospecting the region's coal deposits.

CHAPTER 6

Along the Chukotka Peninsula Coast

The *Iakut* is found—Lemashevskii falls ill—In Nome again—Unpleasantness once more—We return to the Arctic Ocean's Russian coastline—To Koliuchinsk Bay and back—The Americans refuse to work

1/14 July. Regardless that four of the Americans have remained in Cape Nome,[1] the unpleasantness aboard the steamer continues. Because four berths had freed up, Karl Ivanovich ordered a reshuffle, making two cabins available so some of the Russians could occupy them. Picard refused being assigned to share with Scranton the younger, saying he didn't want to be alone in a cabin with a peasant. Al'fred Markovich offered to transfer him to Landfield. Landfield mouthed off to him. "You," he said, "don't have any right to allocate Roberts's and White's cabins, and in general it's time this foolishness stops, otherwise you'll have to deal with Roberts and that won't be good for you."

This was really out of line. In San Francisco this Landfield was a real-life Molchalin, and Roberts asked, then threatened, to have him taken to the bridge.[2] Our Americans are a remarkable sight. Every evening they can all

---

1. According to Bogdanovich, these Americans were Roberts, White, Dowlen, and Scranton Sr.
2. Apparently, to be jailed. Aleksei S. Molchalin is the scheming central character in Aleksandr S. Griboedov's comedy *Woe from Wit* (*Gore ot uma*) (1824).

be seen playing poker together, even the Stuarts will sit in: the Stuarts shake hands; Scranton senior collegially claps Roberts on the shoulder and asks, "How ya doin', George?" But now Picard is refusing to sleep in the same cabin with Scranton, saying he's a peasant.

On 10 July, we arrived at the Miagkov moorage in the Seniavin Strait. Although there was a lot of red and black sediment, Miagkov's gold was not to be found there. The next day, parties were kitted out and departed—the Russians for the mainland and the Americans for Arak Island—but they didn't find gold. For myself, I grabbed a seine and left to go fishing in a lagoon on the mouth of the Marich River, but unfortunately there were absolutely no fish. This was because of the quantity of birds. A flock of ducks was floating on the lagoon; and in the tundra there was the steady moan of avian voices. The flock of birds was small and completely tame—you threw a rock, and they would take wing then settle down a few feet away. There were whole swarms of gnats that bit terribly: Major Stern really swelled up completely.

Some Chukchi were herding reindeer to the shore, and we bought several head. The price was astonishing: they'll accept two bricks of tea for a reindeer, and a brick costs around fifty kopeks. For the same price, you can buy a kukhlianka—a jerkin made from reindeer hide. The tea made available here is generally highly valued, and the words "brick of tea" are practically the only Russian phrase the Chukchi know, pronouncing it "breek of tea." Besides this, there's just "shuker," by which they mean sugar.

The reindeer we bought were killed and skinned immediately. The Chukchi did this as follows: having chosen the best reindeer, they lassoed its antlers, pulled the reindeer to the ground, and stabbed it in the heart with a small, narrow blade: it was a rather unpleasant scene. The reindeer was absolutely terrified, its kind eyes circling as if begging for mercy, and then, when the knife entered its heart, it shuddered, dug its hooves into the ground, rolled its eyes back in its head, and, having stretched itself out, died. They immediately stripped off its hide and gutted it; the blood gushed onto the ground, inundating the reindeer's body, the stripped hide, and the innards.

At 12:00 o'clock we weighed anchor and passed into Kon'iam Bay.[3] This bay is long, deep, surrounded by tall mountains, and makes for a beautiful stop for ships. We made many probes, and in one spot in a stream, at the bottom of which there was a lot of quartz, we found evidence of gold.

I was sitting stuffing a crow, and because birds decompose rapidly, the air in my cabin had turned worse than unpleasant and Lemashevskii said it was reaching all the way to his cabin.

---

3. Kon'iam Bay is today called Penkegnei Bay.

FIGURE 6.1. Aboleshev Bay

In the evening we passed into Aboleshev Bay, in the heart of which is a Chukchi hamlet of three yurts. Bypassing it, we dropped anchor, and next morning went ashore to prospect for gold, but no gold was found. (The captain said that that night, at 3 o'clock, he saw smoke as if from a steamer in the Seniavin Strait. It might have been the *Iakut*, but we couldn't move and just let loose some steam.) After prospecting, we returned to the steamer, but lunch proved to have been eaten. If the whole party was onshore, we asked ourselves, for whom was the meal prepared? It seemed for the ship's crew only.

Today, for no reason, the young Stuart struck a Chukcha in the face; she began crying and complained. The blow earned a rebuke, but he got angry and responded that the reprimand was undeserved. Later, in the evening, Picard returned from shore; he brought with him a small polar fox he captured in the mountains; he wants to raise it and bring it to America. He luckily found an indication of gold, albeit minimal.

We'd spent an entire day filling up with fresh water. Today, the water pouring into our kettles suddenly proved to be completely salty. The senior mechanic [Kinney] blamed Reiner for drawing saltwater, but during a test from the sloop the water proved to be fresh. It turned out that Kinney himself had opened a tap and let in the saltwater. We had to stop for three hours because of this.

FIGURE 6.2. Between Mechigmen and St. Lawrence Bay

Toward evening, we reached Cape Chaplin and noticed that a ship was also traveling there. This proved to be the *Progress*: it was arriving from St. Lawrence Bay, where it had picked up its party, and it brought us news that the *Iakut* has been in Lawrence Bay since 9 July. The *Progress* handed us a letter from the *Iakut*'s commander, Mr. N., saying he'd arrived in Providence Bay at the designated time and is awaiting us. We're finally getting our Russian laborers! We immediately turned around and headed north to St. Lawrence Bay.

9/22 July. I've not written in my diary for an entire week. Something's broken out and I'm losing my head—first the *Iakut* was to blame, but then Lemashevskii got sick.

On the morning of the 15th we entered Lawrence Bay, and the matter was not without curiosity. The captain didn't recognize Cape Novosil'tsev, mistook Mechigmen for St. Lawrence Bay, and dropped anchor south of Cape Novosil'tsev. The problem was explained to him the next morning and he set off for St. Lawrence Bay, but encountering ice floes there, he immediately wanted to drop anchor. It took great effort to persuade him to move to Litke Island, where we spotted the awaiting ship.

There was indeed a great deal of ice, but it was thin and we carefully pushed between the floes to Litke Island. There stood the *Iakut* and the whaling schoo-

FIGURE 6.3.   Sailors of the *Iakut*

ner *Korvin*. We were welcomed aboard the *Iakut* and very kindly invited to lunch. After the American concoctions aboard the *Samoa*, we ate the cabbage soup with kasha and the salmon belly with such pleasure. In fact, the almost daily soups of canned oysters and condensed milk have become so disgusting that to think of them is offensive. Indeed, so nice and friendly were the officers, and so amicable was the assembled company, that we spent the night aboard the *Iakut*.

There is unpleasantness aboard the *Samoa*. Old man Lemashevskii has taken ill. The matter began with a trifle: he was bitten by a mosquito, scratched the afflicted area, and then powdered his hairbrush with soap, this brush having been earlier used to coax our puppy along deck. A small dot formed in place of the bite, but the next day a swelling tumor appeared on his elbow and his temperature was quickly rising. Pus manifested inside the tumor and I and the *Iakut*'s physician, Mr. G., made an incision; this drained the pus, but the inflammation spread higher and higher, his temperature rose, and the situation has turned serious. The patient is 65 years old, and this poses a danger.

It seems the unpleasant altercations aboard our *Samoa* will never end. On the 17th we took on the three sailors, five Cossacks, and twelve laborers from the *Iakut*. Because of this, Captain Jahnsen was very unhappy. I don't know if it was out of displeasure or habit that he began drinking. He drunkenly went on the bridge and, noticing a Cossack Karl Ivanovich had asked to stand guard at Lemashevskii's room, began dressing him down. The Cossack wouldn't budge. Then the captain exchanged fisticuffs with Vasilii, who had gone over to the ruckus, and a battle erupted between them. Even before this, the captain had had a fight with one of his sailors—the captain battling with an oar, the sailor a board—and they beat each other sufficiently well. All this was not enough for the captain: seeing that the Cossack wouldn't leave the bridge and he had no control over him, he grabbed a revolver and threw it at the Cossack. He was apprehended, dragged to his cabin, and locked inside, and his assistant was assigned to guard his cabin door. Such is our captain, and such is the discipline possible! Next day, the captain was sobbing and begging Karl Ivanovich's forgiveness and vowed on his Masonic bible not to drink anymore.

Another scandal now got out of hand—on the night of the 18th to the 19th the *Iakut* left, and we moved to Cape Nun'iamo and sent Miagkov to prospect with a party along the coast. On the 20th it was decided to drop Landfield, Picard, and Scranton's son, and a group of the Chinese, into the middle of St. Lawrence Bay. Nikolai, who is in charge of distributing provisions and tools, had gathered everything the party needed and laid it on deck. Landfield demanded he be given yet another tent and tools on top of those prescribed by Karl Ivanovich. Nikolai refused to do so without an order from Karl Ivanovich, so Landfield instructed Scranton to enter the hold and take what he needed himself. Nikolai blocked the door. Then Landfield ordered: "Hit him." Scranton punched Nikolai in the face with all his might, so that he bloodied his cheek and knocked him off his feet. Nikolai, having lost his wits from the pain and insult, grabbed an axe, but folks rushed into the fray and disarmed Nikolai. Revolvers were now glinting in the hands of the Americans. That's how far the matter got. Karl Ivanovich called on the captain and Stern, and recommended measures be taken so that such disorders not be repeated. Later, the expedition's Russian members gathered in Karl Ivanovich's cabin so as to sort through the incident. A majority decided to make Landfield apologize and to warn him to control himself from hereon, and to leave Scranton's son at Cape Nome. I alone was of the opinion that both should be left in Nome.

We entered the heart of St. Lawrence Bay in the evening and found the little steam schooner *Vera* there. We sent the Cossacks' commander to ask what the schooner was doing. Its commander came to the *Samoa* and declared they were gold prospectors and had come here not knowing of V–ii's concession rights. Karl Ivanovich, having produced documents, asked them to leave. They yielded, but with considerable ire. According to them, they were incurring 15 thousand in losses. The schooner left, though we—having disembarked Landfield and Picard with the party of Chinese—followed after them. There was still a lot of ice in the bay, and while behind Litke Island we had to frequently change locations to avoid the floes.

We made for Mechigmen Bay, which is now ice free; the entrance to the bay is very tight between the flats converging from shore. It was risky to enter, so the sloop was sent ahead to take soundings. From behind the sloop, we gingerly proceeded between the flats, but then the captain turned south for some reason and got the steamer stuck on a shoal. We tried to back sternward to get off the shoal, but we've proven to be stuck fast and have decided to spend the night on the shoal.

11/24 July. We weren't able to get off the shoal until after lunch yesterday, by pulling on two anchors from the stern. We passed through the heart of the bay. In the evening our traveling companion fell sick: fever and a cough. I have incised Lemashevskii every day, but his tumor is rising higher, his temperature is very high, and my hope is growing less and less.

Today I managed to take the time to get out to shore. It was so charming there: the quiet, the warmth, the grasses, the flowers, the sea gently lapping the beach, and farther on, mountain after mountain, neither terrible nor black, but delicate and covered with the blue haze of a light brume. Butterflies flitted and bumblebees droned amid the grass. It didn't at all seem we're just below the Arctic Circle. Today was the warmest we've recorded: the air temperature in the shade was 22° C and the water was 12° C.

On our third day and evening we bartered for a great lot of things from some Chukchi. Now there's a holiday on our street. Earlier, the Americans managed to buy everything of interest right out from under our noses, but now we have a certain Cossack—Commander Tret'iakov—who knows how to speak Chukchi well, so we can communicate with the Chukchi. Through him the Chukchi informed us there's a coal bed on Mechigmen Bay's southern shore, not far from its entrance, and that they've made some use of it. If true, this alone might yield a return. The trip from here to Cape Nome takes a little over a full day, but a *tonne*[4] of coal there fetches no less than 25 dollars and sometimes, toward winter, this price doubles or trebles. The Chukchi said the seam lies close to the sea; if so, then it's no worse than the gold. However, gold didn't turn up there, though according to a sample from the bottom of the bay, the strata here have shifted decisively and there may be gold further on.

---

4. Akif'ëv's transliteration of the French—though in English, now rather archaic—spelling, *tonne* presumably signifies here a metric ton, which weighs slightly different from an American/British ton.

FIGURE 6.4. Digging at Mechigmen Bay

Today, Miagkov asked to be seconded a small, light party to go from Mechigmen Bay to the Koliuchinsk's upper inlet,[5] but Karl Ivanovich refused. For their part, Rickard and Stern are already saying it's time to go to Cape Nome for Roberts.

13/26 July. Yesterday evening, a bit west of Cape Novosil'tsev, we disembarked Miagkov with a party and ourselves passed through Seniavin Strait, to the Marich River to take on fresh water. Not having a purifier on the ship is terribly inconvenient, though it's good there exist many convenient places on our coast for getting water. There was still a great deal of thick ice on the shore where we disembarked Miagkov. Karl Ivanovich inspected the shoreline and found the strata weren't bad.

We reached the Marich River this morning but didn't take on water because the wind was pushing a lot of saltwater into the lagoon. To not waste time, we headed to Kon'iam Bay and there disembarked another party of laborers under Petrov's leadership.[6]

---

5. In other words, overland to the head of Koliuchinsk Bay, on the Chukotka Peninsula's north coast.

6. Petrov is described elsewhere in the diary as the Russians' "senior laborer." He was among the group transferred from the *Iakut*.

It was a beautiful day. The scent of the grasses and flowers variously carpeting the beach filled the air. Snow stood on the mountains, and they were abounding in patches of fresh, bright-green grasses. Birds were twittering in the tundra. After its long, ten-month sleep beneath a cold white blanket, nature has awakened and it was smiling invitingly toward the sun, having aroused it with its passionate kisses. You somehow came to life and tried to breathe in as much air as you could. Onshore was a brook, transparent as crystal and burbling gleefully between the rocks; a small, clean tent was set up.

I could have stayed there in that tent with much pleasure instead of returning to the repulsive *Samoa*, but awaiting me was Lemashevskii, on whom I yesterday made the latest incision, nearly from the shoulder to the elbow, there being nowhere else to incise, but he isn't at all better, and an erysipelatous spot has appeared on the left side of his chest.[7]

From Kon'iam Bay we again went to the Marich, and this time loaded the water. Some Chukchi paid us a visit and one of them, the former acquaintance I'd given a rinse for his mouth, now heartily shook my hand and showed off his healthy tongue and gums with joy. How little is needed to win over a fellow, but what a pity the Chukchi utterly lack the ability to help themselves. It's as if the advance of evolution has completely forgotten them. Nothing indeed can enlighten this godforsaken place. Only if gold is discovered. Then a mass of people might come here and the Chukchi get a more cultured life and certain material benefits. Example, of course, will change the Chukchi, and indeed the Chukchi in Endogai village understood what cleanliness means and the need to wash your person, as I've recalled above. In addition to accepting with great suspicion the advice of more civilized people, many consulted me and the *Iakut*'s ship physician, G–v, but we gave them medicine sadly knowing that all our measures amount to only the slightest palliative, that we are powerless before such capable enemies as the severe environment and complete lack of culture. What we'd seen to that point was sufficient to prove their living conditions are completely unsanitary. Their eyelids and eyes' connective membranes are chronically inflamed by the soot from their yurts' illumination, and cataracts are no rarity. Their habit of sleeping side by side on the floor without clothes and pressed tightly together enables the transmission of skin diseases; also, because of their lack of cleanliness and their unhygienic clothes, skin diseases are very common and obviously transfer startlingly easily. The same may be said of syphilis. Though no one came to me with this illness, it's rumored to be widespread here and, moreover, I found in the

---

7. This is a bacterial infection of the skin and lymphatic blood vessels.

Providence Bay cemetery two skulls profoundly showing all the indications of syphilis.

As for childhood diseases, there's a basis for suspecting measles is here. The sharp changes in temperature, the stuffy, soot-filled atmosphere of the bedrooms with their various miasmas, and the entranceways with that strong wind from outside leading to that terrible, severe frost summon forth the host of bronchitides and pneumonias I saw in Providence Bay, where everyone without exception was sick and coughing. Their diet is poor, of course, they don't stock up sufficient supplies, and each spring they experience chronic starvation; no wonder, therefore, that scurvy and various disorders of the digestive organs are encountered among them.

It's strange that Russians almost never visit the Chukotka Peninsula, even for commercial goals; they really consider the Americans suited to carry on the international trade here. On the contrary, there should be no hindrance to the government turning its attention here and dispatching ships, if only to protect the coast from foreign commercial exploitation. It's like everybody forgot about our distant Russian backwoods and no one's doing anything about it. Opposite, on the American coast, we see a completely different attitude. Customs cruisers ply the roads from beginning to end there, guarding the coast. Why doesn't Russia institute such vessels?

We passed from the Marich River through the strait to Cape Chaplin, and met along the way the American schooner *Anaconda*.[8] Having asked, we learned it was coming out of Cape Nome to barter with the Chukchi. Here was confirmation of what I've just said. From Cape Chaplin we set a course for Cape Nome.

17/30 July. Everything about our arrival in Nome was an utterly trying nightmare. We'd hardly dropped anchor when White and Stern showed up with Rickard and invited Karl Ivanovich to the captain's suite. With his opening words, White began laying down the conditions on which Roberts agreed to travel. First, he insisted on going to the Arctic Ocean, since he knew gold to be there. Second, he himself wanted to command the Americans and the Chinese. Finally, he demanded he be brought to Nome upon conclusion of the Arctic Ocean expedition and that the day of his return be precisely established. Karl Ivanovich agreed to the first two conditions but point-blank refused the third and did so beautifully, since events would later show it was best not to come to Nome. Those events were the scandals our Captain Jahnsen perpetrated daily. On the first day, he invited an absolutely suspect lady aboard the *Samoa* and they dined together in the company cabin; that evening, having escorted her

---

8. "Anaconda" is misspelled as "Anonkonda."

ashore, he returned completely drunk and began shouting to the whole steamer that he wasn't a dog and wouldn't work through the holiday, that he didn't want to continue the expedition and wanted to stay in Nome, and for there to be a new captain. He woke Al'fred Markovich and me up, tore into the Cossack guard on the bridge again, and wouldn't calm down for a long time.

The next day, he gave in and brought the ship a bit south of Nome; returning to Nome, he rushed the arrival terribly and narrowly averted a crash. I was sitting in my cabin when suddenly it felt like the engine stopped. I ran on deck—we proved to have been going full steam toward shore and turned away with only several *sazheni* left to go. Another minute, and we would have wrecked. In the evening, he again brought his lady to the steamer and she stayed all night, and in the morning, having eaten and drunk, he ceremoniously placed her aboard a steamer bound for San Francisco. In sum, he was never ashamed. Nome's residents were laughing at us because there was always a bunch of bottles floating around the *Samoa*. In a word, he'd converted our steamer into a disreputable saloon.

But that wasn't all. Today, he went ashore to the customs house and claimed the Russian sailors and Cossacks aboard the *Samoa*, who are armed with rifles and pistols, snuck them aboard ship without his permission or knowing about it. He returned accompanied by a customs official who informed Karl Ivanovich that, according to American law, armed foreigners are not allowed in their waters and that, based on the captain's statement about not wanting armed foreigners on the boat, he was asking to be given all the weapons. Karl Ivanovich told him the Russian government had loaned out the soldiers to defend the concession's interests and so the weapons were needed, that the ship is chartered for an expedition along Russian shores and, accordingly, Russians may have weapons there. The captain then simply refused to leave and advised Karl Ivanovich to find another captain, to whom he'd turn over the vessel.

"Let the captain put his refusal in writing," Karl Ivanovich said.

"I don't want to," replied Jahnsen.

"In that case, I request the customs official be a witness to your refusal."

"But I will refuse only in case you're dissatisfied with me. Then I'll drop both anchors and wait for whoever my boss Ganeff sends."[9]

Seeing that Jahnsen was drunk, and wanting to end the conversation, the official told him:

"You're now in American waters and you can take the Russians' weapons, or I'll entrust this to someone."

---

9. First mention of this name. Ganeff presumably owned the *Samoa*. Curiously, "Ganeff" is probably a transliteration of "Танев," which name suggests a Russian.

"But the doctor here can take them himself, I'll trust him," Jahnsen said.

"Well, that's outstanding," the official concluded, and turning aside to Al'fred Markovich, he whispered to him: "I can see your captain's drunk, and I recommend you leave immediately so he doesn't cause another scandal."

That evening, everyone agreed to raise anchor. Then commenced the finale. I'd only just finished bandaging Lemashevskii, and he was falling asleep, when suddenly I heard our *Samoa* whistling at full strength. I ran to the bridge, asking the captain not to signal that way so as not to wake the patient, and Karl Ivanovich was asking as well, and White and Stern came running, but the captain continued to salute every steamer without hesitation.

"Here on the bridge, I'm the master! I know what I'm doing," he was shouting.

Karl Ivanovich, green with hatred, came running onto the bridge.

"For the last time, stop signaling."

"Get out of here, I'm the master," answered the captain.

"*God damn*, b . . . , b . . . , listen for the last time to what I'm telling you," Karl Ivanovich, beside himself, was shouting.

"I'm the master here, I do what I want. Stop! Drop anchor!" bellowed the captain.

We were about to collide at that moment with another ship's stern; we turned aside, a couple of *sazheni* from its gunwale. The command came from the top, but the captain's order wasn't followed and the anchor wasn't dropped. White and Stern took the captain to his cabin and calmed him, but having to remain on the bridge was poor Reiner, who had to do both his and his captain's job, and for this he gets a salary of all of 90 dollars a month. This is simply painful in comparison to the salaries our Americans are getting: for example, Dowlen gets a thousand dollars a month, Landfield 500 dollars, Scranton the carpenter and Picard around 250, and even the Scranton boy gets around 100 dollars a month.

At last, we've finally left Nome and are now going to Cape York, to look at the mines. Everyone's mood is very low. Up to now the business has been a failure; relations with the Americans are impossibly strained; God knows what the captain's doing; and to top it all off, the sick Lemashevskii, whose high temperature is still not better, has become terribly enervated and capricious—how will this all end?

22 July/5 August. Circumstances are improving, Lemashevskii has gotten better, the swelling has lessened, and his temperature is lowering. Once again, there's hope he will survive, and this is giving us great satisfaction amid our unpleasant situation. We've sailed a lot these past days. From Cape Nome we went to Cape York, and from there we passed through the Bering Strait to St. Lawrence Bay. We couldn't enter the bay the whole night owing to the fog, and only on the morning of 1 August did we come to a stop in the middle of the bay, so Landfield could take his crew out. Landfield found no gold there, though his report contained some value. They began working the surfaces' edges at no deeper than five feet, but even then seemed more occupied with stuffing a deer. Having picked up the crew, we went to Cape Novosil′tsev for Miagkov. That day Roberts, rinsing sand in the area where Miagkov was, found two granules of gold. This was quite enough for the Americans to get all worked up. Stern went running to Karl Ivanovich in his cabin; his face, always so noble, had turned avaricious and he was out of breath, spitting out words like peas.

"Gold, gold, Mr. B., Roberts found gold!"

"Where did he find it?"

"In the sand, where Mr. Miagkov was left."

We ran on deck and all the Americans were already gathered there, looking through a magnifier at some sand in a saucer.

"There it is, there it is—gold. Roberts just looked at the sand and said: 'There it is,' and he was right."

Two small granules of gold truly could be seen in the sand.

"Quick, to Miagkov," the Americans were urging.

We headed full steam through the bay; but barely had we left than our *Samoa* began bouncing around like a ball. Fog was all around, and the wind picked up. We reached Miagkov's station toward evening. The captain went on deck.

"There's a strong wind and a high surf and going in the sloop will be dangerous, we'll wait till tomorrow," he said.

Almost immediately following his words, we watched as a sloop that had departed shore was diving into the waves—first disappearing completely, then flying on the crest of a wave as if completely divorced from it—and coming toward us.

Half an hour later Miagkov, sunburned and reddened from the wind, arrived on deck.

"There's no gold," his voice resounded. The Americans wouldn't trust him and all left for shore.

The next day, everybody went and even took for themselves the auger, Roberts's celebrated and proverbial auger about which so much had been said. They left for shore after lunch, and we, having gone through the fog awhile, hoisted a solitarily flapping Russian flag on the shore. The Americans proved incapable of finding gold, and the auger—Roberts's pride—proved so unsuited to the task that it was tossed away in disgust on the beach: they'd actually paid 400 dollars for it. Having looked at still another beach, we returned to the steamer and, after talking it over, decided to do the following: leave Nikolai with half the party at that spot, but put Miagkov and the other half in the boat and go north. The sea was absolutely clear, not a single ice floe to be seen.

Having taken Miagkov and the workers aboard and provisioned Nikolai, we raised anchor and went to Cape Litke. There was a dense fog in the morning, hiding everything from view. Calculating that the shore should be to the west of us, we set off in two sloops to investigate. Once we got away from the steamer and found ourselves amid the thick fog, a milky white blanket encircled the boat and left us a horizon of several *sazheni*, and to me it seemed we were cut off from the world more than ever. If the compass were lying we might be going not toward shore, but out to sea, and we'd find ourselves who-knew-where for a long time. And if a storm arose, it wasn't certain we'd make it. Especially since we had, against all regulations, brought no food or water. And so it transpired: the compass began spinning around from the tossing and uneven rowing. It was a good thing we remembered that we had small hand compasses in our pockets; they helped us, and after some time moving forward, we heard the muffled noise of breakers, and then there appeared gigantic, dark coastal cliffs descending steeply into the sea. Bursting from them were swarms of sharply crying birds circling above our heads and frightening us with their calls. We turned north and hurriedly got beneath the cliffs. The cliffs ended after two versts, and we stopped at a gently sloping shoreline, at the mouth of a small stream. Red granite granules covered the whole beach but revealed no gold after being rinsed, though we couldn't hammer the prospecting shaft deep because we didn't even have a simple pump with us. A rising wind drove the fog away, the sun rose, and before our eyes an amazing scene revealed itself. The bleak, greenish sea was speeding its cold waves toward the mighty, straight cliffs and smashing at their feet into a fine white spray, and farther past Cape Litke could be seen a small, peaceful inlet surrounded by a string of green mountains. Flocks of tufted puffins, cormorants, ducks, and various gulls were flying around the cliffs.

The next day (4 Aug.) we went a little further north, beneath Cape Dezhnëv (Cape East), and two parties, Russian and American, went off to prospect. It was foggy and cold, a fresh wind was blowing, and it was getting choppier. Around noon the captain, saying not a word, raised anchor and proceeded a mile north. It was already around 3 o'clock, but our miners still hadn't returned, indeed, since they now couldn't find the *Samoa* because it had changed location, and the fog was such that nothing could be seen past ten *sazheni*. I asked the captain to blow the siren: we began issuing one desperate call after another.

FIGURE 6.5. Cape Dezhnëv

At four o'clock the sloop with the Americans appeared, coming toward the siren, but there was nothing of ours. We began to worry, for it wouldn't be hard for them to get lost. Around 6 o'clock, the sloop finally appeared off the gunwale opposite the shore. So they really had gotten lost: coming back from shore, they'd charted a return course for the steamer but of course couldn't find it. Two hours had passed and there was nothing of it; they were getting tired. What was there left to do? You can't find a ship in the fog, and they'd have to return to shore and wait there until it cleared, wait without a crust of bread and without a stick of wood to heat some water. (Nearly all the trees have been washed away from the coasts here. One gazes enviously at the American beaches, at all their bracken.) They were floating backwards, utterly downcast, but then the wind blew and Karl Ivanovich saw the ship's mast in the distance. They hurriedly raised anchor, and came alongside half an hour later, worn out and hungry, since they'd not had a poppy seed in their mouths since morning. For all that, they brought good news: gold was found in every chute rinse, albeit very little.

That night we rounded Cape Dezhnëv. We were proceeding not far from the shore, encased in fog. The wind was terrible, and our unfortunate *Samoa* was rocking mercilessly. Now and then the captain halted the steamer and dropped a sounding weight. The depth was always around 30 *sazheni*. During these stops the wind would turn and carry the steamer, and finally, the captain could not guess where we were. The wind was very strong, and anchoring at sea was impossible, so we had to find Cape Dezhnëv and gain cover behind it. We moved on and nearly grounded on the cape itself: we were very lucky the fog parted and we saw before us the cape's enormous cliffs and heard the breakers' roar. We immediately reversed course and began rounding the cape, keeping farther from it. Toward morning the choppiness grew, and a wave washed away the whaleboat dangling from our sloop davit; we had to look for it in the dark, though the night was even darker now; however, we found it despite the choppiness, though upon raising and landing it on deck, we discovered there were only four oars.

Our entire circumvention took fourteen hours. But this morning, when the fog broke, we proved to have gone ten miles to the north and had to reverse course. We dropped anchor opposite the village of Uélen,[10] located beneath Cape Dezhnëv's north side. This village consists of several yurts. Its Chukchi have a strong and healthy appearance and evidently live well, more or less: this was visible in the volume of their yurts and of the goods they brought to our ship. The Chukchi here have absolutely none of the characteristics of the Mongol people. They are tall and solid; their eyes are cut straight; their skulls do not protrude at all; their noses are oblong, albeit with narrow bridges; the skin on their faces is bronzed and dark, though the skin on their bodies is like that of the other Chukchi living to the south. Some spoke English, though poorly, and none, Russian. Only one greeted us, saying: "Hullo [*Zdorovo*]." Americans evidently visit them frequently, since they had whaleboats and Winchester rifles in sufficient quantity.

The Chukchi crowded aboard our ship the whole day, exchanging their goods. Without haggling, I gave what they asked; I purchased five polar fox furs for 6 bricks of tea and four large walrus tusks for 4 bricks; speaking for myself, this was terribly cheap, and it's no wonder that American whalers don't pass up opportunities to barter with the Chukchi. Most interesting of all, we bought a huge polar bear fur. The Chukcha man wanted a Winchester .44-caliber rifle for it: we offered him the rifle, a cover sack, and a box of bandolier cartridges, but the Chukcha didn't take it, saying the rifle was too small and light and that he needed a heavy long gun. He haggled for a long time, and we even had to go ashore, but he finally agreed to give up the fur for 30 bricks of tea—i.e., nearly half as much as we'd offered him. In essence, there is in their trading no correct price at all: they accept what they need at a given moment, and often at far less cost than they propose. Their requirements are

---

10. Akif'ëv spells "Uélen" as "Uéllen."

FIGURE 6.6. Uélen drydock

sometimes amusing. After being given various furs, belts, and this and that, they cut off nearly all the buttons on one of our sailor's jackets.

Next, we purchased an entire collection of figurines expertly carved from walrus tusk. The figurines are shaped like different animals: deer, dogs, foxes, seals, the heads of walruses, birds, and even entire dog teams with sleds, and all this for some two or three bricks of tea, a handful of beads, and some poor knives and similarly cheap items. Our Americans traded widely. Ignoring what we've often told them of the prohibition against trading in liquor, they often cunningly and unnoticeably did so, bringing the Chukchi to their cabins or going ashore and trading the goods in the yurts. Our female traveling companion noticed White entertaining Chukchi with whiskey onshore and how Reiner was carrying something similar inside a small *sac-voyage*.[11] This young assistant was yet cleverer. I saw him handing over a rubber sack for hot water used in place of a warm compress, but he was using it to convey whiskey to the shore.

We tried searching for gold: we hammered a prospecting shaft 6 feet deep, but there proved to be neither gold nor substrates.

---

11. The female traveling companion referred to is Bogdanovich's wife.

23 July/6 August. Today we moved several miles north and stopped at the mouth of a large lagoon, counting on supplying ourselves with fresh water. Karl Ivanovich and Miagkov went ashore early in the morning for prospecting, though the jolly company of Americans left at noon. After some while, strange shooting was heard from shore—this was them working. Strictly speaking, they nearly always work like this: toward noon they collect themselves, and bring with them an abundant supply of wine, beer, snacks, and rifles. Onshore, they first of all go hunting and only then engage in prospecting, though they don't go deeper than three or four feet before turning back.

It proved impossible to supply ourselves with water. A strong wind arose and the captain, having taken over the cutter, got it stuck on a shoal in the mouth of the lagoon: to dislodge it from the shoal took great effort. Within the lagoon itself, our men nearly overturned and just barely got away. Toward evening we raised anchor and went further north. There was almost no fresh water on the ship, and what we had was brackish. We needed to go straight to Koliuchinsk Bay, where there are streams.

Passing near shore, the captain noticed water falling from the cliffs and we stopped the ship.

"This is where we can take on water," he rejoiced.

The water was falling from a high cliff into the sea and looked from the outset impossible to get to. I expressed my opinion, but the captain replied:

"I've traveled the California coast sixteen years and quite often got water from waterfalls."

I finished giving my opinion, and even bet a pair of half-dollars with his assistant that they wouldn't get the water. After supper, we decided to reconnoiter the area; nearly all of us went; from a height of 10 *sazheni* the water was falling heavily onto a stone slab, near which was just a narrow strip of sand. Gaining purchase on that amid the breakers was going to be hard, but notwithstanding the strength with which it was falling they still wanted to go after the water. The captain turned obstinate.

"I said I'd get it, and I'll get it. Tomorrow morning we'll connect the chutes together, put them under the fall, and fill up the sloops."

We had to give in and spent the night at anchor. Nights have now turned long and dark, and fog surrounds us with showers drumming the deck. The water temperature is already just 34° F.

28 July/11 August. We're stuck in Koliuchinsk Bay, waiting for good weather so as to get water from here and go south. For the past two days a strong SW wind has been blowing and it's been impossible even to release a sloop. This evening the wind died, and at midnight Miagkov, so as not to lose time, left for shore aboard the cutter.

In my diary I've established how we were off the coast, planning to take on water. Given that we now have to wait, that of course proved impossible.

At seven o'clock in the morning[12] the wind picked up, but the captain, who had been lost in his charts until practically midnight, only got up at nine o'clock. They readied the chutes for sluicing the water, coupling, positioning, and laying them out; they came back four hours later.

"Well, how much did you get?" we asked.

"It was impossible, the power of the fall wouldn't allow it," answered the assistant.

"Good God, how much good water do we have now?" Miagkov, having come on deck, turned to the captain.

The captain muttered something and went to the bridge. We raised anchor and headed north, having lost some 18 hours thanks to the captain's obstinance.

Toward evening we rounded Cape Sredtse-Kamen.[13] It has a very unique appearance: individual rocks cover its entire peak like stubble, giving the appearance of turrets. Past Sredtse-Kamen we ended up in a dense fog, owing to which we couldn't find the entrance to Koliuchinsk Bay and had to spend the night anchored in the open sea; but the wind greatly strengthened and the choppiness turned very strong. Next day, the fog began thinning and we moved closer to shore, but we still didn't enter the bay straightaway. The entrance to it proved to be not at all close to what was shown on the charts. On all the charts the entrance looks several miles wide, but in actual fact the entrance is all of two or three miles across. It was a good thing we didn't risk going at night and that there wasn't a storm, for without a doubt we'd have crashed into the shore.

It's odd that this entrance alone is shown as extremely wide on all the charts. Indeed, a century ago there was a Russian expedition here, under the leadership of Captain Billings, and in the accounts of this compiled by Captain Sarychev, it says on p. 30: "The mouth of the Koliuchinsk Inlet is 4 miles wide" (from the report by Sergeant Gilev).[14] Though some information was later given the

---

12. Of 24 July / 7 August.

13. Translatable as "Cape Heart-Stone," this is located on the Chukotka Peninsula's north coast.

14. The Englishman Joseph J. Billings (1758–1806), a participant in Cook's third global circumnavigation, entered Russian service in 1783. Under orders from Catherine the Great, Billings explored the Chukotka Peninsula's interior during the winter of 1791–92. It was apparently this expedition that resulted in the coastal charts Akif'ëv refers to. Lieutenant Gavriil Andreevich Sarychev (1763–1831) served under Billings. His account of his expeditions with Billings was published in 1794, and it has recently been republished as Gavriil Andreevich Sarychev, *Puteshestviia flota kapitana Sarycheva po Vostochnoi chasti Sibiri, Ledovitnomu moriu i Vostochnomu okeanu* (Moscow: Izdatel'stvo "E," 2016).

government, everyone obviously ignored this piece of news. Nordenskiöld,[15] who wintered in these parts, investigated the mouth of Koliuchinsk Bay during winter when it was covered with ice and snow, and was obviously unable to notice the two long sandspits extending from west and east that narrow the bay's entrance. The spits are indeed very narrow, and it's no wonder they can't be seen in the snow. They're covered by scraggly vegetation and a bunch of small freshwater lakes, from which we provisioned ourselves.

In the evening we went ashore to go hunting and, with four rifles, killed around two hundred sandpipers. There's such a crowd here you can simply aim a rifle wherever and they don't even scare off. Of the four types of sandpipers here, one is very interesting: the spoonbill, with a nose that flattens to a small oval scoop.[16] A bunch of forest bracken covered the spits' beaches, and in spots whole piles of tree trunks lay. In this regard, the Arctic Ocean's shores are much more promising than those of the Bering Sea and Strait, and the tides don't strip them clean. Signs of previous Chukchi habitation were visible on the spits, but no new villages were spotted anywhere. We didn't find gold on the spits, though Landfield did have the good fortune to find three gold granules that appeared in the chute. The permafrost begins a quarter to half below the surface here.[17]

It was impossible to take on water yesterday: the wind was blowing strong from the northwest and the temperature suddenly dropped low. Waves were picking up in the bay. Our *Samoa* was rocking from side to side as if on the open sea. The scene turned gloomy: turbid-green waves were making a noise against the gunwale as if grumbling against being raised by the wind; the sky was completely swaddled in clouds, and a wall of ash-gray fog stood around us; the wind was pouring cold showers onto the deck, whistling savagely, hissing and pulling at the tackle and irately banging doors and hammering windows. Were you to go on deck, it would immediately whip your topcoat up and try to pull your hat off.

This morning the wind began showering us with snow—this is the first snowfall. Toward evening, when the wind had quieted and the fog lifted, we saw the mountains were covered in snow. We need to get ready to leave here, because the temperature keeps falling and the wind's blowing from the southwest and may soon bring ice, which always builds up near land at Vrangel.[18]

---

15. The Finnish Swedish scientist and explorer Nils Adolf Erik Nordenskiöld (1832–1901) led the 1878–79 Vega Expedition, which involved the first complete crossing of the Northwest Passage.

16. Akif'ëv calls this bird the "kulik-kolpik." It was evidently a heron of the genus *Platalea*.

17. It is not clear what unit of measurement Akif'ëv is using here.

18. More popularly known as Wrangel Island, this was named after Baron Ferdinand von Wrangel (fon Vrangel'), a Baltic German serving in the Imperial Russian Navy.

Nordenskiöld stopped to winter there because he hadn't noticed how the wind brings the ice, and his path to the Bering Strait got cut off.

Nevertheless, we have not spent time here in vain. Miagkov followed the bay's entrance to the nearest shoreline, completed measurements, and essentially corrected the map so that the bay can now be entered without danger. The bay is very good for laying over: it's wide and so long that its ends cannot be seen; its fairway is 15 or more *sazheni* deep and—from the spit—runs two hundred *sazheni* along the western shoreline (though it's shallower along the eastern). Small mountains with conical peaks surround the bay, taller mountains are visible to the southwest.

6/19 August. We're stopped in the small Seniavin Strait, supplying ourselves with fresh meat and water and cleaning the engine. We left Koliuchinsk Bay on the 11th, just as the wind died down. Passing Sredtse-Kamen, we saw a herd of several dozen head of walruses; we could even see their enormous tusks, stretching 14 *vershki* long at times. This has been the first and only occasion we've seen walruses. We weren't at Sredtse-Kamen long, though Karl Ivanovich inspected the shore in the meantime; and the following day there was another stop at the village of Nuukan, situated in a small square amid the crags on Cape Dezhnëv itself. We were suddenly surrounded by a little fleet of sloops loaded with furs. The trading was ferocious and we bought a mass of furs; furs can sometimes be purchased surprisingly cheaply. I was able to buy a polar bear hide for 10 bricks of tea, 8 pounds of sugar, and a hunting knife. The hide was outstanding, 4 *arshiny* long from head to tail and with something of a dense, yellowish hair; Lemashevskii (now fully recovered, to our universal relief) traded 18 bricks of tea for this same hide. It's a good thing that a given deerskin usually costs 1 brick of tea, the same as a fox or a polar fox fur. In general the affordability is terrific, and Americans take advantage of it to the full extent, bringing with them for any opportunity whiskey, which can buy anything here. Were Russians to make up for this, it would be to the benefit of both them and the Chukchi.

Having finished trading, we headed south, to a strip of beach beneath Cape Dezhnëv where we'd found some gold. Karl Ivanovich had, the evening before, ordered Picard to go to this beach with a party of Chinese, and Al'fred Markovich got all the party's provisions ready that evening. As we reached the designated spot, Picard said:

"Mister Roberts forbids me to go ashore."

"Then let Picard write me that Roberts forbade him to obey me," said Karl Ivanovich.

Picard left and came back several minutes later.

"Roberts forbids me to put anything in writing."

"Then you'll verbally repeat this before witnesses."

"Very well, I agree." And he repeated it.

And so the Americans are straightforwardly refusing to work and refusing to obey Karl Ivanovich. What was to be done? You can't find a labor force, you can't lose time, and there's little that remains, so the work had to be done immediately. Karl Ivanovich ordered all the Russians and all the Chinese to get ready for work.

After a little while, the Chinese boss—i.e., the labor crew's contractor—appeared in Karl Ivanovich's cabin, and told him:

"We were hired to work for Mister Roberts, and as soon as Mister Roberts orders us to stop working, we stop, because we consider him our master."

In such way the Americans and the Chinese were refusing to work, and we were left with only three Russian laborers. Karl Ivanovich then sent the sailors and the Cossacks to work, and they, without a murmur and with evident pleasure, went ashore. By a necessarily general comparison, our Russian laborers and soldiers are not just distinguished by their discipline but are even straightforwardly interested in working, and they have a desire to work. It's a pity not all the laborers are our own: Nikolai's and Petrov's two parties still haven't been picked up, but had they been with us then, we wouldn't have been especially aggrieved by the rebuff from the Chinese and Americans, since none of them will ever be as good as our workers.

We all went ashore. We picked up a rock, pounded in the prospecting shaft, and got to work. Karl Ivanovich stayed with the shaft and Miagkov and I went to inspect some nearby streams, he to the south along the shore, me to the north. The geographical strata in the streams turned out to be very good. The entire bottom of a stream consisted of quartz, micaceous schist, and gneiss, and if there's gold in these streams' mouths, then it should be imagined there's even more of it in their upper reaches.

Paying no attention to the strike by the greater half of the expedition, the Russians continued working in this area for two days until it became undoubtedly clear that gold really is there, and we took several poods of sand for rinsing in Petersburg.

Roberts and company had had to stay in the same spot, but because they considered the expedition finished, they were demanding to be taken to Cape Nome and put up there. Karl Ivanovich answered that he wasn't going to Cape Nome: first, because we still had to pick up Nikolai's and Petrov's parties and could not leave them onshore, and second, because there's little time left and we still have to inspect the peninsula's entire south coast from Providence Bay to the Anadyr River, and therefore can't waste time traveling to

Nome. The Americans said this was infringing on their liberty and that they would demand of their captain, as their fellow citizen, to immediately deliver them to American shores. To this, Karl Ivanovich responded that they, as members of the expedition, had to submit to his authority, that he saw no infringement of their liberty, and that the captain, by force of contract, also had to submit to him.

Then the Americans shut up.

From Cape Dezhnëv we went to Nikolai's party waiting beneath Cape Novosil′tsev, and picked them up. The party had worked diligently, hammering prospecting shafts 26 quarters deep, but hadn't found gold. From there we went to the inlet of Kon′iam Bay, for Petrov's party, which had now been sitting on the beach nearly three weeks. They, too, had not found gold, despite hammering twice during their time up to 30 stakes, some as deep as 22 quarters.[19]

We'd now collected everybody, and it had become possible to prospect the southern shoreline. We stopped to take on water, venison, and to clean the engine, which hadn't been cleaned for a month. So as not to waste a single day, we followed the small Seniavin Strait and stopped near the island of Shurluk, opposite a third, still unnamed, bay. The workers were divided into three parties and sent ashore; also, the Chinese finally gave in, and several went to work.

The Americans stand stubbornly on their own. They don't live bad, though. They drink, eat, play cards, and for diversions, go ashore to hunt, stroll, and trade with the Chukchi and, taking advantage of our absences, have already purchased several chests of furs. Today, they went to Shurluk Island and Dowlen traded away his rifles—given him for the expedition—for furs.

Visiting me in my cabin for conversation, they're quite amiable. Especially White. He's always sharing how delighted he is by our surroundings.

"It's outstanding, it's staggeringly beautiful, just like in Switzerland!" he exclaims.

It really is very nice around here and somewhat reminiscent of the Swiss lakes, though of course not on such a grand scale.

We now have a clean engine and a supply of water and have purchased meat. At 9 o'clock this evening, we raised anchor and gingerly proceeded through the strait so that we'll be in Providence Bay tomorrow morning.

---

19. As above, it is unclear what distance a "quarter" refers to.

CHAPTER 7

Our Arrest in Nome

Being arrested as pirates—The trial

7/20 August. We're heading full steam to Cape Nome. Yesterday we all stretched out peacefully in our cabins and slept, planning to wake up in Providence Bay. Awaking this morning, I could feel the steamer was going all out. What did this mean? Indeed, given the passage of time, we should already have been in the bay. I grabbed my compass, and it turned out we were going east. Why that way? To get to Cape Nome, of course. I bounded onto the deck: there was water all around and no shoreline visible. So it was, they were taking us to Nome.

Entering the stateroom, I encountered Miagkov.

"Did you know we're going to Nome?" he said.

"I know, I saw by my compass."

"Really, the devil knows what's going on. This truly is coercion! I won't be offering them my hand after this."

Rickard entered at that moment.

"Good morning, Mr. Miagkoff!"

Miagkov turned his back on him, and Rickard left the table.

Having gulped down a cup of coffee, I went on deck and ran into Al'fred Markovich.

"Did you know we're going to Cape Nome?"

"Can that be?!" he lost his wits and even rubbed his eyes.

"I'm telling you the truth."

"What does this mean?"

"We'll find out later."

Karl Ivanovich approached.

"After you've tied things up, everyone come to my cabin, and we'll assess our situation."

We gathered ourselves together. Miagkov began first.

"Karl Ivanovich, allow us to stop their diversion; really, power is on our side: we have lots of people and guns."

"No, that's impossible. They may not heed the threat and it could lead to bloodshed. I've no doubt we'd take the summit, but a lot of blood would spray and we'd have to answer for it. We'd be taken to court and then, believe me, we'd get blamed for everything and no one would stand up for us. For now it's better to lose the grudge and end this affair diplomatically. If we behave correctly, then there's no doubt we can exit this unpleasant situation with honor."

Al'fred Markovich went up top to ask the captain, who was standing on the bridge surrounded by the entire horde of Americans, to come provide an explanation. They proved to have been carrying on all night there. The captain responded he wouldn't come and turned his back on him.

Karl Ivanovich then dispatched a Cossack with a letter to him—Jahnsen wouldn't accept the letter and sent him back. Landfield appeared in the cabin.

"The captain wishes to speak with you, Karl Ivanovich, in his cabin!" he said.

"I want an official response from him via his assistant," Karl Ivanovich responded. The senior assistant appeared several minutes later and said:

"The captain wishes to speak with Bogdanovich in his cabin, in the presence of Al'fred Markovich and one other Russian witness, during which he will also have a single witness."

"No," said Karl Ivanovich, "that isn't convenient for me—let him come here."

Reiner returned once more.

"The captain says that henceforth, he does not consider you the commander of this expedition but only a passenger and therefore, if you like, he will speak with you in his cabin."

Nothing happened, and we sat writing a statement about their removing us from the Russian coastline by force. Just as we finished, White and Reiner appeared in the cabin. Both looked awful, White was actually trembling as if he had a fever.

"What explains your visit?" Karl Ivanovich asked them.

"We've come in the captain's name."

"In that case I've nothing to tell you, the captain's refused to speak with me."

"If so, we've come to speak with you on our behalf."

"Then that's another matter. Please sit down."

"The captain says he's going to Nome so the American authorities there can explain to him who Bogdanovich is and how we should treat him."

"First, the captain knows that according to the contract I'm master of the ship during the term of the lease; second, I yesterday informed him in writing that I'm a Russian official appointed by the government and the expedition's commander, for which I have official documentation. I presently consider myself to be under arrest and am unable to account for this situation."

They retreated, but quickly returned again.

"We and the captain request you give your word that the Russians will not do anything by force. We're familiar with Russian discipline and will trust your word."

"If I'm no longer acknowledged as the commander then I cannot give orders. Furthermore, should any one of you touch a Russian, I won't deny him the possibility of acting within his rights."

"Nevertheless, the captain, as master of the ship, asks you to take measures," said White, trembling and utterly pale as snow.

"I consider myself a prisoner and therefore am unable to take any measures. Let the captain himself issue the orders, but I myself refuse to enforce American laws."

Having thus achieved nothing, they withdrew. Then, sucking down a beer, I took one of the workers' rifles and laid it next to me in the cabin we were sitting in, to avoid any misunderstanding.

The steamer is traveling at full speed, such that everyone's bodies are shaking. The Americans haven't quit the bridge, they're all armed with revolvers, and the junior assistant is even carrying his in a holster belt. Well, so now we're prisoners. It's obviously the end of the expedition and the beginning of all sorts of wounds and afflictions. And all the while we've endured much from them.

8/21 August. Cape Nome. I'm sitting in my cabin admiring a guard's gleaming bayonet—we've been taken prisoner. Strictly speaking, this ain't bad for the start of the 20th century. We sit as captured pirates; here I am, not having anticipated being given such a notorious title and landing in such a situation. Our arrest transpired as follows.

We reached Nome yesterday evening and dropped anchor. A doctor came, went through the steamer, and left. After him, Reiner went ashore and in a little while returned, grabbed some documents, and went off once more. No one met with him, and no one came to us aboard the steamer. The Americans didn't sleep the entire night: the sailors were posted on deck as watchmen, whereas the gentlemen sat in the captain's cabin. Even there, they were frightened.

This morning they were again expecting some customs official, but in vain; no longer waiting, they went after lunch to a government hydrogeographic ship that was stopped in the roads; the captain took himself there grandiosely but, though taken onboard, was not invited inside. He produced some document but evidently achieved nothing, since he returned to the *Samoa* empty-handed as the hydrogeographic ship turned around and left. He then steered the *Samoa* toward an American barracks located several versts outside Nome and went ashore with Roberts in a sloop.

I laid down to rest and, having awoken, was watching through the window. I saw a sloop with ten soldiers and officers moor alongside; the soldiers came on deck, rifles loaded, bayonets fixed. Not bad for starters. Following them there moored, along with the captain, an elderly gray-haired officer in a faded-red canvas jacket; he went to the captain's cabin and invited Karl Ivanovich and Al'fred Markovich there. This officer is the local military commander and a major.

"Why are there armed Russian soldiers aboard the *Samoa*?" he asked Karl Ivanovich.

"The Russian government gave us soldiers for guarding the interests of a Russian expedition, and all of us have been brought here by force," answered Karl Ivanovich.

"Why did you, captain, leave the Russian coast without an order from Bogdanovich?"

"Roberts, Landfield, Picard, and Scranton informed me in writing that they feared for their safety and demanded to be taken to Nome immediately. One of the Russian laborers, speaking in English, confided that the Russians wanted to commandeer the ship on the way to Anadyr and leave the Americans onshore. Therefore, I immediately brought them here."

"I have to get the soldiers to the Russian naval ship *Iakut*, which will be leaving Indian Point soon, so I request the captain return me there," said Karl Ivanovich.

"Indeed, the soldiers need to be brought there," agreed the major.

The captain straightforwardly refused.

"I'll go only if I'm given 8 soldiers as a guard, or if a military ship escorts me."

"I could resolve this myself, but I prefer to confer with the civil authorities," the major concluded, and he left.

Toward evening a military vessel arrived, and two officers came, sniffed around, and went away saying they'd return the next day. The vessel departed.

The Americans launched the cutter and took it to the city, but we've remained under guard. The situation is strange: wherever you look, you see bayonets.

This evening Miagkov went into the living compartments, grabbed all our cartridges and, with sack in hand, was returning to the quarterdeck: something glistened before his eyes and a bayonet poked his chest. The sentinel yelled and up sprang soldiers who surrounded Miagkov.

"What's the meaning of this? Karl Ivanovich, please come here!" shouted Miagkov.

Everyone went on deck. The matter was explained and Miagkov released. It turns out the major has given instructions to block communications between the quarterdeck and forecastle at night, and he has, moreover, forbidden us entry to the stateroom via the deck, rather than directly along the interior stairwell. As such, we're being denied freedom of movement on our ship, this being done in the Land of Freedom. It was enough for some rogue like Landfield or Scranton to say the word and all thirty foreigners have been arrested and put under guard.

9/22 August. Today Karl Ivanovich sent the major a letter, in which he asked for an explanation as to why, exactly, soldiers have been posted over us on the ship and when they'll be removed. The answer was that the soldiers have been posted to maintain order and the Russians should not consider themselves under arrest, since we can realize complete freedom when it is convenient to go ashore, and that he was even inviting us to visit him in the barracks for tea.

We were grateful for his curious invitation, but do not at all agree that we're completely free. We can't take a stroll or speak with the laborers during the evening, and all day long, the soldiers with their rifles are walking around and even bringing their rifles into the lavatory. They sit in the stateroom eating and drinking beer and wine, and sleep on the floor. This evening, our young officer's wife came and visited us on the steamer as well, but immediately got seasick even though there weren't any waves.[1]

"Well, I wouldn't want to sail in your ship," she said.

Indeed, everyone's moods are worse than usual.

We got a newspaper. There was an article about us:

"30 Savage Russian Cossacks Have Been Taken Prisoner by Seven Brave Americans" ran the headline in large letters.

---

1. This was apparently the wife of the American officer in charge of guarding the Russians.

Russian Cossacks (this article reads), totaling 30 men under the leadership of Dudenedich (so his [Bogdanovich's] name was transcribed), wanted to bring the vessel *Samoa* deep into Siberia and slaughter or abandon them there on a deserted shore and commandeer the ship.

The Americans, having overheard this talk, waited for a convenient opportunity, and such presented itself.

One night, when all the Russians were sleeping, the Americans occupied all the best strategic spots, armed themselves with revolvers, and diverted the vessel toward Nome. The Russians, having awoken the next day and realized the change of course, flew into a frenzy. But the seven brave armed Americans managed to restore order among the thirty Cossacks and to subdue them. The vessel *Samoa* is now under arrest and the Cossacks are awaiting trial.[2]

In such way, we are captured pirates. The Americans' actions have been dreadfully solemnized. All this led to the *Samoa*'s sailors fancying themselves to be my masters in full: they threw my boxes, bottles, leaves of Circassian tobacco, and the birds I'd stuffed into the water. My heart simply bled watching this. The captain calmly ordered the cutter up and managed to smash it into the stern and to peel its side. To Karl Ivanovich's question as to whether he had the right to seize someone else's property, he answered that the cutter was listed on the ship's manifest and so he considered it as belonging to his ship. The major refused to interfere in this matter. As such, our property has been plundered, no one wants to defend us, and we remain under arrest.

12/25 August. The soldiers were on our ship for three days. Finally, the *Mining*, a military vessel, arrived.[3] The major and Jahnsen went to visit the *Mining*'s captain, and the three of them reached a decision on this business. Unfortunately for us, the *Mining*'s captain was a friend of Jahnsen's and a fellow Mason. The major relayed their decision to Karl Ivanovich. It was as follows:

Captain Jahnsen would bring the sailors, Cossacks, and laborers to the *Iakut* off our coast, then return to Nome and take us to San Francisco.

Such a resolution could not be entertained, of course, and Karl Ivanovich decided to bring the matter to the customs administration.

Yesterday, the soldiers were taken off the *Samoa*, and the ship was moved to the commercial road and anchored between two military vessels.

---

2. I have been unable to find this article or anything similar to it in the Library of Congress's American newspaper database.

3. No record of anything even approximately called the USN *Mining* has been found. Akif'ëv seems to have been mistaken regarding either this ship's name, or provenance, or both.

FIGURE 7.1. American guard aboard the *Samoa*

You glance to the right, there's cannon looking at you, to the left, cannon as well. It's very nice to feel so safely guarded while strolling the deck.

This morning, Al'fred Markovich went to Nome with Karl Ivanovich's statement for the customs administration. There, the customs agent Colonel Evans and his assistant Captain Jarvis, an old mariner who once commanded the cruiser *Bear*, accepted it.[4] As Al'fred Markovich explained the matter, they became very interested and indignant at the military leadership's behavior. Jarvis came to us and spoke with Karl Ivanovich aboard the *Samoa*, then the two of them went to see Evans.

Around 12 midnight we observed the Northern Lights. The Cossack on watch came to me and said:

"There's a fiery glow above the city."

---

4. Built in Scotland in 1874, the *Bear* was a forerunner of modern-day ice breakers. It entered American service as a revenue cutter for the Coast Guard in 1885 to become one of this branch's most famous ships. After being acquired by Canada and renamed the *Arctic Bear*, it sank in 1963. Akif'ëv is mistaken in identifying David H. Jarvis as the *Bear*'s former commander: Jarvis's relationship to it was as its first lieutenant, when in 1897 he won fame for leading an expedition of *Bear* crewmen to save several vessels that were stranded in ice near Nelson Island, in the Bering Strait. See Preston Jones, *Empire's Edge: American Society in Nome, Alaska, 1898–1934* (Fairbanks: University of Alaska Press, 2007), 38, 52; SS, USRC, USCGC & USS Bear (defense.gov), accessed 26 April 2022; Mission Plan: Search for the U.S. Revenue Cutter Bear: NOAA Office of Ocean Exploration and Research, accessed 26 April 2022.

I ran on deck. Visible above the city to the north was a phosphorescent blue; the light grew brighter and brighter. An arc formed, spanning a quarter of the surrounding horizon, and rays were shooting up from it; then another arc began forming beneath it and this phenomenal scene lasted nearly an hour. The light was blue, quite powerful, and shimmered slightly, like a wet phosphor candle burning in the dark. Towards the end, spaces appeared in the arcs and the light began to disappear.

14/27 August. There's a terrible storm. The barometer's falling fast. The sky's clouded with fog and curtained in gray, hiding the city from us; the sea froths and drives forth huge, deep-rumbling, white-crested waves. We're rocking hard, and the other ships no less than we. Here in front of us, the wind is pulling on some white steamer like it's Noah's ark. We haven't raised the anchor and it's drifting, first turning the gunwale and then the bow toward a wave. Now it's drifting again. The wind's getting stronger and stronger. This storm's relentless whistling and howling, this fog, dampness, rocking, all of it is producing a terrible despondency, a desire not to do anything.

We've been here at Nome for a week.

Karl Ivanovich and Al'fred Markovich went ashore yesterday morning and returned at six o'clock in the evening. A meeting took place where, in addition to our own people, Roberts with his lawyer, and Jahnsen, Ricker, and Landfield participated. Who could have predicted that Ricker, who alone among our entire company we considered a most decent fellow, is now relying on a lawyer? From the start of the meeting, the captain was so flustered by the lawyer's questions that he blathered all sorts of nonsense about Evans and Jarvis already providing a solution.

During this, the lawyer said:

"This matter is beyond your competencies, none of you have the right to decide it."

They absolutely lost their minds over this. It was completely obvious that Jahnsen is a scoundrel and was up to dirty tricks, but he can't be punished or forced to fulfill his obligations. Yet that wasn't his only lie: he said of me, that I threatened him on the bridge with an axe; of Nikolai, that he'd wanted to murder Scranton; and that Roberts had announced to him that if he didn't turn the ship around things would get unpleasant and then he, Roberts, would take over and kill Bogdanovich. This was all written down in a testimonial.

In general, everyone blathered a lot; so nothing got done last night and it was left till today, but now there's a storm and going to shore is impossible.

Yesterday, I read in the newspaper that there was a meeting on the 25th aboard the *Mining*, where it was all Americans and not a single Russian. It was decided in this meeting that Jahnsen was right. Not a bad meeting: they question just one side and decide in their favor. Evans and Jarvis acknowledged that yesterday's meeting was illegal and rejected its decision. A pity there's a storm today, for the matter could probably have moved forward.

I've forgotten to say that a certain Podgurskii, or a *Count* Podgurskii, as he's proclaimed here, came to visit us. He initially offered his services through a letter, then he himself arrived, talked with us for a long time, grew exasperated with our Americans' behavior, and finally concluded thusly: "But, you know, Karl Ivanovich, it's nevertheless better if you make peace with the Americans and not initiate a case. I'm sure it will turn out bad for you." In a word, it seemed to me he'd been sent by Roberts and company to parley with Karl Ivanovich.

Prices here ain't bad: Al'fred Markovich paid 11 dollars for a sloop from the shore to the steamer and back—i.e., 22 rubles for an hour's work.

Manners here also ain't bad: they're gossiping about the following incident. A certain gentleman climbed atop another during the night: the sleeper woke up. The designated guest then drew a revolver and said to the occupant:

"Don't fuss, sir, lie quiet and I won't touch you."

He lay there until the guest took his watch and money and went away.

The victim got up, went to the window, and noticed the robber entering one of the taverns. Then he pocketed his revolver, went into this tavern, saw the robber at the bar, and, pointing the revolver at his head, also politely said:

"Don't fuss, sir, stay quiet and wait until the police arrive."

The police came and arrested the thief at 5 o'clock in the morning. His trial was held at nine o'clock that same evening and the verdict delivered: 10 years' hard labor.

Such are manners and practices here.

16/29 August. We've finally been freed from imprisonment; we've suffered not a few sundry unpleasantries. The expedition is now ruined, of course, and we crave merely for a quicker transfer to the *Iakut*. They finally decided this morning what to do with us.

Our entire company of Americans was placed onshore, and old Colonel Evans was supervising them. A pity it wasn't Jarvis, since he's smarter and more energetic, for this old man, though educated and courteous, is mentally enfeebled. He soon made one tactical error. Just before departure, Jarvis noticed him with two ladies.

"Where are you planning to go, Colonel?" he asked.

"Well, I want the pleasure of a lady and we're going for a spin to the Siberian coast."

With great difficulty, Jarvis persuaded him that given the circumstances, it was extremely inconvenient to leave with the ladies.

We left at 6 o'clock in the evening. The captain quickly grew familiar with Evans, and they were soon drinking champagne in his cabin. Evans was exchanging bows and smiling like a child and, as Karl Ivanovich said, was focusing all his attention on keeping his dentures from falling out of his mouth.

As could be expected, Captain Jahnsen appeared before Karl Ivanovich this morning, repented of himself, cursed the Americans for putting him in a spot and then renouncing him, and said that Bogdanovich was a most decent man, the most intelligent and refined in all of America, and told him that if only they didn't consider the contract broken, he would have gone to Anadyr or the Strait of the Cross or anywhere that was suitable.[5] He also showered praise on me to Evans. He'd made himself a laughingstock by evening.

Behind us there's a puff of smoke in the distance. A military vessel can be seen with binoculars to be following us. We aren't bereft of escorts after all.

---

5. The Strait of the Cross is in the Gulf of Anadyr.

CHAPTER 8

Release from Captivity

Return to Russian shores—The end of the expedition—Aboard the military transport *Iakut*—Petropavlovsk, Kamchatka

18/31 August. Here we are in Russia, though nothing resembles our European Russia. We're surrounded by wilderness and barrens, the silence is deathly, these cliffs are naked, and the sound is a bit quieter than that which hung over Nome with its noisy throng of various rascals, wanderers, swindlers, and seekers of easy profit.

We reached Indian Point yesterday before evening and went ashore and bought some things. We gifted Evans with walrus tusks and a bird-skin kukhlianka and he was as delighted as a child. I gave him several photographs of the Chukotka Peninsula: he put them carefully away and said he'd bring them to his daughter.

Today, we're in Plover Bay. We were late waking up when Miagkov came running.

"Do you want to go to the Olenna River?"

"Of course I want to, because after ten days' captivity on ship, my legs feel like they're petrified. Moreover, the boredom's terrible and there's nothing to do at all. I've already packed our collections that were spared from the plunder, and I've laid up the pharmacy. Lemashevskii is almost healthy and has only three small wounds in his arm."

I got dressed, grabbed a rifle, and settled into the cutter. How nice it was not to see a single American with us. Only Evans was sitting there, but he had asked so sweetly to be taken ashore, to get Siberian flowers for his daughter, and was so childishly delighted by everything, that the only impression he made was a good one. We traveled the entire day along the length of the Olenna River, and only late in the evening did we finish and return to the steamer.

30 August/12 September. We're sailing the Bering Sea aboard the *Iakut* and are now approximating Petropavlovsk. These past days are as if in a fog.

While at Emma Bay, where the *Iakut* arrived and we transferred to it, there was no time to write. First, they took up our case once more, and again there were questions and again they had to rummage, so to speak, through our dirty laundry and then transfer our stuff.

As soon as the *Iakut* arrived, Karl Ivanovich went to it in good form and officially informed N–ii [the ship's captain] of the foregoing unpleasantries. There were indignation and threats, more it seemed of the latter; among the officers as well. For them to come from Petropavlovsk to the Chukotka Peninsula was unpleasant enough, but knowing they had to complete such an outing yet again, they'd become terribly quarrelsome. Also, they had to investigate the clashes and unpleasantries.

Second, they aren't particular enthusiasts for long journeys, and this undertaking hasn't been easy. For example, there are only three officers of the watch aboard the *Iakut*. Every day, each has to stand watch for 8 hours. This is quite fatiguing. Just imagine if some predatory ship has to be stopped. How can this be done? It has to be staffed with its own officer command, but there aren't enough officers. It would be impossible to transfer one or especially two to the ship. Aboard our *Samoa* there were three, since the captain himself stood watch during the day, and they were terribly fatigued, so our ship isn't going to catch predators at all. What can they do with a ship, just seize the documents? It will just move on and then, in some distant port, change its color, change its name, and travel wherever possible once more. In a word, protection of our coastline and Russian interests in the North does not in fact exist. A single small military transport gets sent to Petropavlovsk, but only once every several years are things surveyed farther north. For this reason, foreigners feel at home along our northern shores. Contrast the United States: military, customs, geodesic, and zoological ships are there. Now, that's protection.

So N–ii undertook to investigate our case. He formed a commission of two officers with himself as chairman. The next day, he invited Jahnsen to appear before him along with Colonel Evans, in the capacity of interrogated witnesses. Jahnsen testified he was always happy with the Russians and saw nothing bad in

FIGURE 8.1.   Olenna Chukcha

them. He'd gone to Nome out of necessity, to avoid disorder aboard ship. As a matter of fact, Roberts had wanted to murder Bogdanovich and he, Jahnsen, twice took his revolver from him. Moreover, the expedition's commander and members demanded he bring them to Nome because they'd heard the Russians wanted to take them deep into the countryside and murder them there. Reiner and Kinney were next summoned, and they testified they were happy with the Russians and that the captain brought the ship to Nome

without their asking him to. The *Samoa*'s sailors who were interrogated testified that relations with the Russian laborers were always good and they never
feared them.

To avoid any misunderstanding, the commission decided to take all the written testimony and turn it over to the local leadership and to transfer us to the
Iakut for removal to Petropavlovsk.

On 26 August we transferred to the *Iakut*.[1] The weather was nasty, there
was a pouring, hard rain, gusty north winds, and fog covering the mountains.
At night it was freezing and ice formed on the water's edge.

Having built a barracks in which to put all our now unneeded tools, the laborers returned from shore before lunchtime. It was a pity to see them tossing
around things that had cost huge amounts of money. The snow will start falling, build to a blizzard, and bury this barracks; in spring, the snow will melt,
water will flow, and everything will rust through. If there is another expedition
these things will be worthless. Contrarily, perhaps the *Samoa* command, having
returned to Nome, will tell of them and some American schooner will snatch
them.

Al'fred Markovich called on me just as I was getting my last things
together.

"Did you know the captain's still dreaming up more trickery? He's demanding that Karl Ivanovich must accompany him to San Francisco."

"Whatever for?"

"He says that per the contract, Karl Ivanovich has to supply the steamer
with provisions and coal, and therefore he either has to give him money or
remain aboard the *Samoa*."

"Well, what did Karl Ivanovich tell him?"

"'You yourselves,' he said, 'invalidated the contract, so I'm not bound to
honor it. I'm certainly not leaving with you, nor am I giving you money.'"

"But the captain groused, 'I'm going to complain to the *Iakut*.'"

"Evans, seeing that a storm was brewing, started crying. 'This drunkard's
gonna drown us,' he was saying, 'so make him stay here 'til tomorrow.' Pity
the old man, for who's going to persuade that drunk Jahnsen? He's already raising the steam."

Walking on deck, I ran into the captain, who could barely stand on his feet.
He grabbed my arm and said:

"I've always considered you a good man and a gentleman."

"But how do you regard Roberts and White?"

---

1. The text gives a date of 26 September here, but this is obviously a mistake given that a few
lines below, Akif'ëv lists the events described as having occurred on 26 August.

"I now realize they're scoundrels," he said. "Everything would have been fine without them. Now I'm going to have to bring them to Nome for nothing, but I'll get some paying passengers and leave for San Francisco."

"That would be good," I said.

"Farewell, Doctor, and I wish you well. Don't forget me."

Reiner approached after him.

"Farewell, Reiner."

He pressed my hand, but his lower lip was trembling, and his eyes turned muddy.

"Don't think poorly of me, Doctor. I'm not a free man and have to submit to the captain. But I always esteemed and considered you to be of good people. For this I suffered gibes every day from Roberts's party. I couldn't share a word with you without them looking sideways at me. Farewell!" As if having become stooped, he departed embarrassed. I was sad that he was the only kind person among them all.

I was instructed to lower the sloop, put the last of my baggage in it, and go with Miagkov to the *Iakut*. The wind was whipping up, the bay turning choppy, and rowing was now difficult.

The others arrived within an hour. After this, the *Samoa* emitted a long, mournful whistle, and having raised anchor, inched slowly away. Shouts were carrying from its gunwale and the Americans were waving their hats and handkerchiefs, and our laborers responded likewise. And why shouldn't they have? They'd neither quarreled nor fallen out with the *Samoa*'s sailors, and for the past two days had even been getting drunk together, stinking drunk, of course. As is customary, our Russians would come to blows among themselves but in morning commence once more to drinking amicably.

Being separated from the *Samoa* is rather strange. There's a terrific storm in the open sea and we haven't much ballast.

We left Emma Bay on 26 August and the Chukotka Peninsula's coastline soon disappeared from view. I felt rather sad. There'd been something attractive in that peacefulness and wilderness.

Our ocean journey is boring, the choppiness always quite strong. I'm unemployed and have tried busying myself with photography, but the premises for this are so narrow and close it's absolutely unthinkable to sit there when it's choppy, and moreover, the senior officers don't like me doing this for some reason. But it's nice that there are books, and for whole days I lie and read. In the evening we play dominoes; and take turns going to breakfast and lunch at the commander's. So it goes, day by day.

The officers of the watch are always tired; they don't like talking. It's odd to see so many superstitions among our officers. For example, whistling is forbidden because the weather will turn bad, and there's a sign forbidding scratching your nails on the table for this same reason. Ask when we arrive at Petropavlovsk and this is a straight-out crime, and everyone turns morose.

"That depends on the weather."

"Well, let's suppose wind and weather are favorable."

"Akh, please stop, there's such a bunch of secondary factors."

You see that answering is scary, so you don't ask. Conversations turn exclusively on maritime life, or lifetime reminiscences are traded, and the conversations are extremely dull.

In sum, it's monotonous and boring. Nevertheless, because among themselves the officers are friendly and impressive folks, you feel much better than aboard the *Samoa*.

31 August/13 September. Petropavlovsk, Kamchatka. In the morning, we saw the outlines of the Kamchatka coast emerge from the fog. Straight ahead towered the snow-capped Povorotsk Volcano, and to the left rose the sheer white cone of the Viliuchinsk Volcano. The fog grew sparser the closer we got to Avachinsk Bay, we finally left the gray wall behind us, and before us the sun was shining in a clear blue sky and illuminating a magical panorama.

The wide bay, which can hold ships from all over the world, is encircled by rings of tall mountains covered with bright green forests and yellow grasses; behind these mountains are the gigantic conical volcanoes, sparkling with an attire of white snow extending to their very peaks; peering from the blue sky were shafts of sunlight that flooded the bay's emerald waters, and green, welcoming forests.

Aboard ship, it was as if everyone had been revived, awakened from deep slumbers. How nice everything is here compared to the Chukotka Peninsula's cold, deathly environment. There's so much life!

After lunch we entered Petropavlovsk Bay and dropped anchor near Signal Point. We were immediately visited by the district's assistant commander, a teacher, and a marine supervisor, and were delivered newspapers. Everyone greedily grabbed an issue. How much news there was inside! The war with China was in full swing and had already claimed victims: several of our fellow officers had been killed, and among the names of the dead was that of my local countryman Colonel Antiukov.[2] Starting off with this news was very grievous.

---

2. Akif'ëv is referring to events transpiring during the Boxer Uprising (1899–1901), which offered a pretext for Russia to invade Manchuria and send troops to Peking (later renamed Beijing) and other cities.

The war has just begun and there are already many victims, and there will be more. The insurgency has seized all of China, and not just separate militias but the Chinese people will combine forces for the struggle. This is a people's war. Till now such a war has prevailed in southern Africa, and the enormous British army can't compete with the handful of Boers who've chosen to sacrifice their lives for freedom. Up to now the Americans have been unable to impose complete control over the Philippines. What will happen if hundreds of millions of Chinese fight with similar passion and ferocity for their freedom? And should a certain animosity emerge against the Chinese for their savagery, then on the other hand you must acknowledge that everyone has a right to defend his home against uninvited guests.

Karl Ivanovich and his wife have left for shore, but since finding accommodations in Petropavlovsk is nearly impossible, our courteous officer-hosts have invited us to stay until a boat comes to the *Iakut*. This evening in the stateroom there was the liveliest discussion, us finding something to talk about again, and the boredom disappeared under the influence of new and vivid impressions.

15/28 September. Tatar Strait. We're traveling north aboard the German commercial freighter *Peyang* direct to Aleksandrovsk Post,[3] within sight of the Sakhalin coast. We ended up waiting in Petropavlovsk an entire eleven days for a steamer.

Petropavlovsk itself was not especially interesting. It's more a village than a city and is not a village of riches. It sits on the north side of Avachinsk Bay, deep inside a small cove that is separated from the bay by a tall promontory called Mount Nikolai, which completely obscures it. It includes all of several dozen small wooden buildings and two churches—a winter one and a summer one. The streets aren't paved and have wood-plank sidewalks. There's never any traffic on the streets, nor is singing or music heard: it's somehow remarkably calm and quiet. This delighted our workmen. They went ashore the next day and commenced to feast. By evening, those drunken miners were lying and crawling about all over the streets and in the weeds. They'd lay themselves right down in the street, stand their vodka bottles beside them, and drink, and having gotten drunk, would sing and drink again until they nearly passed out on the spot.

It's impossible to get around Petropavlovsk at night. There's no lighting whatsoever, and so walking along the street you every now and then stumble upon either sleeping cows or drowsing miners. None of them move and no

---

3. Aleksandrovsk Post was Sakhalin's capital at the time.

one pays them any mind; so pleased were they by this lethargy, they erected and stayed in two tents on the beach like they were in a dacha.

The following facts are enough to demonstrate how quiet and patriarchally [*patriarkhal'no*] simple life is there.

There's a gunpowder magazine on Mount Nikolai. This is a reminder of Petropavlovsk's previous greatness when it was still a military port. A Cossack watchman without any weapons was standing at the magazine, which holds practically nothing. I was witness to the following conversation.

A woman approached the Cossack.

"Ivan, go eat!"

The Cossack was counting the time: he wanted very much to get off duty and go eat. Suddenly, he noticed another Cossack walking along the shore.

"Grigor'its!" he shouted. "Oh, Gri-gor'-its!"

"Whaddya want?" he responded (they weren't speaking in hushed tones).

"Come take my post for me, I wanna go eat."

"My cabbage soup's gettin' cold," he immediately answered, and walked off.

"And *my* soup's gettin' cold!" the watchman concluded, and he went off to eat.

I saw him after lunch, sweetly relaxing in the sun in front of the gunpowder magazine, and beside him sat the woman whispering something.

Here's another vignette. At 12 o'clock I was walking past the building where the police station and cooler are located, and where the assistant district commander lives, and I saw the Cossack watchman, this time in uniform and even with a saber, quietly ringing the doorbell to enter the house to eat. He was small and rather quiet and bashful, and so the Cossack saber he carried looked big on him. We greeted each other and began talking.

"Geez, what a handsome saber you have!" I said. "Show it to me."

Together with its scabbard he pulled it out for me.

"But extract it!"

"But there ain't no pulling it out!" the warrior surprisingly and rather sheepishly answered. The two of us took to tugging on it, I from the handle, he from the scabbard, but we couldn't extract it. Clearly, no one had deigned to pull it out for many, many years.

Finally, here's another instance. One of the miners was drinking, carousing, and brawling with his mates. Well, it got too much, and they stuck him in the cooler. Fëdor was sitting behind the grille, enviously watching his mates tethered to their vodka bottles and crawling over fallen tree bark through a green meadow. A Cossack guard was standing in the doorway, observing this scene with benevolent curiosity.

With unbelievable powers, one miner managed to drag himself over to the window.

"Fëdor! Oh, Fëdor!"

"What?" gloomily answered the captive.

"Why're ya sittin' there, brother?"

"Well, I'm just sittin'!"

For several minutes there was an awkward silence.

"But leave!" the visitor impatiently advised.

Fëdor looked and was silent. He watched the sunlight bathing the green meadow with its miners lolling about higgledy-piggledy like cows, he saw the bottlenecks extruding coquettishly and provocatively from his mates' pockets, and he couldn't restrain himself: he banged on the grille and began breaking it. The rusted grille, behind which no one had probably sat for many years, fell apart, and the prisoner crawled to freedom.

The guard stood calmly watching.

"Welp, he's gone!"

And Fëdor sank into the grass as well.

"You wanna drink, brother?" a mate extended a bottle to him.

Fëdor tipped his head back, closed his eyes, and sucked on the bottleneck.

Several minutes later, the bottle empty, the fugitive prisoner was sleeping peacefully with his mates beneath the prison window, but all the same, the Cossack was gazing at them peacefully and affectionately.

The assistant district commander came outside.

"What's this?" he asked, seeing the smashed grille. "Where's the prisoner?"

"Right there!" the Cossack calmly pointed to the sleeper.

"Phooey!" And waving his hand, the assistant went to his place.

These "Kamtsatka Cossacks," as they call themselves, were so kind and friendly. I was able to learn so much about these very honorable, dependable, and peaceful people's expeditions. But they possess no military prowess, and no one would say they are the progeny of heroes. Among themselves, all evince a mixture of Russians with Kamchadals. They're short, homely, and poorly proportioned, but regardless of this I retain a fond memory of them.

However small and bad Petropavlovsk is, it contains some interesting monuments. There are monuments to the seafarers Bering and La Pérouse.[4] On the tip of the sandspit in the bay there's a monument to the victory in '54, during the war; there's a mass grave of warriors beneath Mount Nikolai; and

---

4. Vitus Jonassen Bering (Ivan I. Bering) (1681–1741) was a Dutchman in Russian service who commanded the First Kamchatka and Great Northern expeditions of the early eighteenth century. The Frenchman Jean-François de Galaup, Comte de La Pérouse (1741–88?), was a Pacific explorer who charted the North Pacific during 1787. While in Oceania the following year, he and his expedition disappeared.

FIGURE 8.2. Petropavlovsk Cossacks

on the opposite side of Avachinsk Bay there's a monument to the British admiral.[5]

The Crimean War was a glorious time in the life of Petropavlovsk. If on the Crimea the Russians suffered horrific casualties, then in Petropavlovsk they got the better of the Allies.

The Anglo-French fleet had passed into Avachinsk Bay and was preparing to take Petropavlovsk. There were three Russian warships in Petropavlovsk Bay at the time. Taking note of the unwelcome fleet's proximity, the Russians prepared their defense. They positioned cannon along the end of the sandspit cutting across Petropavlovsk Bay, and off this spit's tip they loaded down their ships with so many rocks that only the decks were visible. When the Allied fleet appeared from behind Signal Promontory and advanced on Petropavlovsk, it encountered a terrible cannonade. It retreated and stayed behind Mount Nikolai. Seeing the city could not be taken from their hiding place, they gathered beneath Mount Nikolai and began quietly ascending it. It need be said that Mount Nikolai offers a gentle, thickly forested incline toward the city and that its perpendicular summit is several dozen *sazheni* above sea level. The

---

5. The victory in 1854 refers to the Crimean War. The British admiral referred to here is Rear Admiral David Price.

Allies summitted the mountain and opened fire on Petropavlovsk, thereby killing 18 residents, but in the meantime didn't notice the Russians stealing upon
them in earnest through the woods. At the appointed bugle call, the Russians
hurled themselves on their enemies on the summit, and they crushed and
drove them below. This business lasted several minutes and during this terrible skirmish 700 men in the Allied army perished. Following this, the Allied
fleet's admiral ordered his ships to the other side of Avachinsk Bay where, unable to endure the shame of defeat, he blew his brains out. The fleet went to
sea and Petropavlovsk was left intact.[6]

Those slain in battle were interred at the base of Mount Nikolai, and erected
above them were three crosses and a small village chapel, where memorial
services are held each year. During our time in Petropavlovsk these somehow
passed unnoticed without making an impression.

The day after our arrival, a ceremonial luncheon was held aboard the *Iakut* to
honor these names, of whom there were three on the ship. For this a tent was
stretched from the stern over the deck and fully decorated underneath with
various national flags. Lunch was served on the deck beneath them. The guests
arrived and even didn't forget a phonograph, and there was a sumptuous feast.
A great lover of music and song, the commander suggested listening to the phonograph. The phonograph played several tunes, then something began to
wheeze. One officer actually suggested pouring wine down the phonograph's
horn. In a word, we so let ourselves go that we celebrated for three whole days.

On 4 September a memorial service was held for Officer Dedenev, who perished in the war.[7] The *Iakut*'s command and officers gathered with several
city residents in the church. An elderly archpriest performed the service. He
could barely walk, spoke very quietly, and kept forgetting everything. For example, he forgot how to invoke the Heir, and recited five or six names before
reaching his.[8] He was muddling his duties, but when this was mentioned to
him, he responded:

"God knows that I know my duties."

There was so much touching certainty in his response.

To somehow interrupt the boredom, we took to hiking either up Mount
Nikolai, where we could relax in the trees' shade, or along the shoreline, to go

---

6. David Price did indeed kill himself, but Akif'ëv grossly inflates the number of Allied dead: all
told, the Anglo-French force suffered 209 casualties, with the number of dead and missing (including
Price) totaling just 52. On the Crimean War's Eastern Theater, see John J. Stephan, "The Crimean
War in the Far East," *Modern Asian Studies* 3, no. 3 (1969): 257–77; and Barry M. Gough, "The Crimean
War in the Pacific: British Strategy and Naval Operations," *Military Affairs* 37, no. 4 (1973): 130–36.

7. Akif'ëv seems to be referring to the recent casualty in China.

8. The "Heir" refers to Nicholas II's son Aleksei Nikolaevich.

target shooting. Once, we repaired to the village of Seroglazka, 5 versts from Petropavlovsk, and found a little spot beneath a tree, ordered a samovar, milk, and cheese curds be brought, and commenced a sumptuous feast. Be it known that everyone got rather wild and behaved like children. We played checkers, and the doctor even squatted and then fired a bullet through his peak cap.

On 8 September the *Iakut* went over to Tar'ia, as the inlet on the opposite side of Avachinsk Bay is called. We then went on land. Our tent was brought and set up on the beach not far from a lake, though we ourselves went off in various directions to go hunting. The hunt was satisfyingly successful: I killed five ducks, but got terribly fatigued, since it was extremely hard to walk there. The grass was so thick your legs got tangled up, and it grows taller than a man, so you had to call to one another so as not to get separated.

Everyone gathered in the evening at the tent, lit a campfire, and began cooking cabbage soup and drank wine and sang until it was ready. The sailors also set up their tent beside ours and listened to the singing from there. Everyone felt light and free, and we even spent the night on the beach. We were to leave Petropavlovsk the next day.[9] We breakfasted with the commander a final time, even drank Champagne, and bid our courteous hosts farewell. When we were exiting from the side of the ship the officers came on deck and the command gathered on the forecastle, and friendly wishes for a safe journey followed after us.

On the 10th the schooner *Kotik* and the Chinese Eastern Railway steamer *Mukden* arrived in Petropavlovsk.[10] We now had to choose which ship to take. The *Mukden* would not be going directly to Vladivostok for another twelve days, but the steamer *Peyang*, of the coal-mining firm Kunst and Albers, had finished loading coal and would leave on the 11th, albeit not to Vladivostok but to Aleksandrovsk, on Sakhalin.[11] Miagkov and I had been wanting very much to visit Sakhalin, so we decided to act thusly: we would go with our laborers aboard the *Peyang* to Aleksandrovsk Post, and Karl Ivanovich and his wife, Al'fred Markovich, and Nikolai would stay to wait for the *Mukden* and join us in Vladivostok. We told the *Peyang*'s captain this, and he agreed to take us.

Two of our laborers have decided to remain in Petropavlovsk. One wants to live it up there but the other, Vasilii Beliaev, was completely smitten with Kamchatka and has resolutely decided to stay there forever. He even went

---

9. In fact, Akif'ëv he did not leave Petropavlovsk until the eleventh.

10. As noted, the Chinese Eastern Railway was Russian-owned and -operated. The *Mukden* would have served primarily to transfer coal and other supplies for use on the railway.

11. Akif'ëv erroneously spells "Albers" as "Alberts." Also, Gustav Kunst's and Gustav Albers's enterprise was not renowned as a "coal-mining firm." It instead consisted primarily of a trading company, the largest department store in Vladivostok, and a store in Korsakovsk Post on Sakhalin. Anything Kunst and Albers did with coal would seem to have accounted for a minor part of their operations.

shopping around Tar'ia and is now fitting himself out as the owner of a small house near the springs. Hence, he's not going home to Kostroma Province, as he earlier planned to. It'll be good if he doesn't start drinking there, as he has great wont to do. Then again, the boredom of the place will probably cause him to get soused.

It must be said the boredom in Petropavlovsk is horrible. It's completely cut off from the world for more than half the year; a telegraph necessary for communication does not exist, there's only the mail, and during winter there's no communication whatsoever until a steamer arrives. The populace exists in their little shells and it's like they go into hibernation. With the arrival of spring, they instantly rouse themselves upon news of the first ship and its arrival. Anxiety grows daily, residents seeking by the minute as to whether the telltale smoke has appeared at the inlet's entrance. The day the first steamer arrives is the biggest holiday of the year. As soon as the steamer is spotted in the distance there's bustling and uproar in the city. School classes are canceled, workshops locked up; everyone hurries to get better dressed and to rush to the shore. By the time the steamer drops anchor in the bay it's like all the Petropavlovskians have spilled out. The jabbering goes on till evening. Residents welcome not just this first steamer with joy: each vessel brings something new and interesting. The Petropavlovskians greet the first steamer with such joy and see the last one off with an equal amount of sorrow. Away it goes, and Petropavlovsk is once more cut off from the world, doomed yet again to a months-long lonely and melancholy existence.

We spent our final evening at a teacher's place. He's already been there several years with his family and had much of interest to say. He spoke of Kamchatka being a land of riches. He himself hasn't toured it, but was told this by a Dr. Tishov, who's studied Kamchatka well. For an entire six years Tishov has completed annual pilgrimages along the peninsula; this year he actually resigned from his position as district physician to be free to go to the mountains. He travels without any money and has accustomed himself to the climate and to the lifestyle of the indigenes, among whom he spends whole months eating dried fish and living in a yurt, just like the indigenes. He is collecting a lot of data and eventually wants to write a fundamental essay. The teacher showed us a piece of quartz with gold flecks in it, a very valuable possession that Tishov gave him. He says that besides gold, there's lots of iron and copper in Kamchatka. There are garnets and amethysts and even pearls in the sea off the Okhotsk coast. There are many forests and many animals in them. The rivers and seas teem with fish. Fishing industries, earlier looked at askance, are now being established. Massive amounts of herring and keta and chinook salmon, etc., are being caught. In essence, were Kamchatka not so distant and cut off from the mainland during winter months, it would attract thousands of people and would thrive.

We returned directly to the *Peyang* late in the evening. Walking along the beach, I was intrigued by the phosphorescent sea: all the sand was phosphorized whenever the water receded. We left bright lights behind us with every footstep. If you spread some sand on your palms they began glowing as if rubbed with phosphor. I found a tiny crab on this beach, no more than a centimeter long, which crept along my hand and was glowing brightly. I stuck it in a glass, and it continued to glow for two days. Even walking along the beach at Cape Dezhnëv, I'd noticed the sea was tossing up bunches of tiny crabs and some sort of thin, transparent worms. I gathered them in a glass and brought them to the ship. One evening I walked into my cabin, and I saw everything in my glass was glowing with bright phosphorescent light. In the Indian Ocean, I'd attempted to capture a glowing creature in the bathtub but it proved nearly invisible to the naked eye.[12]

The *Iakut* left on the morning of 11 September; we went down for a farewell visit to it, and then the officers paid us a visit; upon drinking a good-bye glass of wine, we left Petropavlovsk at 11 o'clock in the morning. We were told not to leave before the new moon but rather during the equinox, but our German captain, a courageous man, said that a ship as good as the *Peyang* can well withstand any storm. Indeed, a strong wind picked up as soon as we left, it became choppy, and fog covered everything. By evening Miagkov and our new traveling companions, the comrade prosecutor Bunge[13] and an agent of the Kunst and Albers firm, were seasick, so I dined with the captain alone.

There was a curious incident that night: the Chinese boy had poured water into a washing cup and left it; during the night, it got very choppy and all the water poured onto the head of Miagkov, who was sleeping on the lower berth near the washing cup. Half asleep, he screamed like a maniac at so undesired a bath.

We've been quiet and bored, day after day. The rocking has been very strong, and our fellow travelers play patience for entire evenings. There was a terrible storm last night and it was so rocky it really was hard to stand on your feet. We will finally hit land today and be able to send our first telegram home concerning our return.

---

12. This reference to the Indian Ocean refers to Akif'ëv's return from his previous visit to Asia during 1898.

13. This may have been Fëdor F. fon-Bunge, who in 1902 was appointed Sakhalin's first vice governor and eventually became Sakhalin's last governor under tsarism.

CHAPTER 9

Sakhalin

Aleksandrovsk and Korsakovsk Posts—Prisons, laborers, administrations, etc.

16/29 September. We approached Aleksandrovsk Post in the morning. A strong wind was blowing and enormous gray-green waves, their white crests breaking and sending cold, salty sea spray into the air, were slamming the shore. An ashen, foggy haze coated everything; we gazed coldly and joylessly through it at the severe coastline of terrible Sakhalin.[1] The steamer made a huge circle and, not getting closer than a mile from shore, dropped anchor. There are no inlets at all here, so it's always bad for ships to stop; when a storm erupts—and they often do here—steamers have go to De-Kastri Bay, on the mainland opposite Aleksandrovsk Post. Our captain was looking around suspiciously and also hinting that we'd probably be unable to go to shore today, but then decided to let us go and blew the whistle to summon a steam cutter. It was very choppy, and it seemed that going to sea in a cutter was now impossible. From the steamer, waves could be seen lashing a small wharf and reaching the top of a little building atop it. We became dejected; but after half an hour, some smoke appeared behind the wharf, and a little later, a small black dot separated from shore and advanced toward us, disappearing now and then behind the waves.

---

1. On the penal colony, see the sources listed in this book's Bibliography.

It seemed like it wasn't getting bigger, but it once again scaled the crest of a wave and, much larger, approximated us. Finally, the cutter was alongside us, but remaining there was impossible, so the cutter crew fastened it to the steamer using boat hooks so as not to slam into it. It was hard to load our baggage and harder still for us to transfer to the cutter. The baggage was put on a hawser and slid down to dangle to a point where it could be grabbed. They were shouting from the cutter: "Faster, faster, we can't hold it." The baggage was finally loaded and stowed away in the cutter, then we had to be transferred. Before we could get into the cutter a sort of complicated gymnastic exercise first had to be performed! A storm ladder was dropped down. We had to descend it, jumping above the water and trying not to slam ourselves into the side of the steamer, biding time for the right moment when the rocking was such that we found ourselves above the cutter, then release our grip and let ourselves drop down, then gather ourselves and stand up on the deck. All this we performed splendidly, and we shouted "Glückliche Reise" [Bon voyage] to our captain and left the steamer. "Hold tight, gentlemen," we were advised, and was this ever so. Immediately, we dove steeply, then a wave washed over, soaking us from head to toe. Then we descended just like a duck. "It's really quite inconvenient to travel in this cutter in such weather," we turned to one of the sailors.

"This is nothing, we've gone to Nikolaevsk[2] in this cutter, but one time a storm blew up and carried us into fog for two whole days. A strange frost coated everything, and finally we were tossed ashore not far from here, it turns out. Still, we were lucky: folks kept life and limb and the cutter wasn't smashed, so it wasn't bad."

So with this courageous seafarer we got under way, forging our frail tub through the stormy Tatar Strait. Yet these brave seamen all proved to be penal laborers or exile-settlers. They got us safely to the wharf and we asked them to return to the steamer for our laborers, promising 10 rubles for this. They gladly reversed the cutter and disappeared once more behind the waves, while for our part we sat ourselves down on the wharf and drank tea beneath the noise of pounding surf.

The first news we heard was not especially pleasant, namely that all the apartments in the social club had been taken by those belonging to the military or to local ships. But not several minutes passed before we learned by telephone that an apartment had been found and horses would be dispatched for us momentarily. This courtesy had been shown us by our traveling companion, comrade prosecutor fon-Bunge.

---

2. Nikolaevsk is on the mainland across the Tatar Strait.

A carriage soon pulled up, and Gennad'ich and I left. We'd been assigned an apartment in the home of a certain exile-settler, the joiner Plessak. This was a large, clean room with beautiful furniture made from tree burls. (Products made from tree burls are a Sakhalin specialty.) Our landlord is an outstanding joiner, has a workshop here, and makes excellent furniture that can be found in Vladivostok and Nikolaevsk. This furniture is very expensive and Plessak does quite well for himself. As soon as we arrived we were given tea with milk, home-baked bread, and sausage, and began snacking with pleasure. In the meantime, Bunge came to see whether we were doing alright. We immediately paid him a visit, and so, too, the local commander, Semëvskii, though we didn't find him at home. We had to find someplace to eat. Our landlord pointed out the community assembly. This assembly is a small club for Aleksandrovsk Post's state servitors; located there are a serviceable kitchen, library, billiards room, and buffet. This building has a small garden, where it's possible to stroll in good weather; and military and civil personnel gather there in the evening to while away the boredom. On holidays a military orchestra plays and there are even fireworks. Being newcomers, we were introduced to the company and spent the entire evening with them.

17/30 September. We spent the morning in the Sakhalin Museum. Sad impressions! The collection is large and in sufficient order, though the building is very crowded. I saw the museum director, Doctor Pogaevskii; he is obviously very interested in the museum but said he has simply let it go, having seen the complete impossibility of doing anything constructive inside such a building.[3] Actually, there would seem to be enough lumber and free labor. But it's said carpentry work is terribly expensive here. Returning from the museum, we encountered the local mining engineer K. and lunched at his place. He came here with his young wife right after their marriage, and, of course, felt it to be savage. We accepted his invitation to show us the mine shafts.

In the evening we dined once more at the club. We were also enlightened as to certain intimacies of Sakhalin life; of course, this was gossip that came after some drinking bouts, though the gossip resembled accusations, and, as later became clear, there's a whole legal case, though so filthy it was hard to listen to.[4] We returned to our residence with an exceedingly heavy feeling.

18 September/1 October. Aleksandrovsk Post. Yesterday we were informed that keeping money in the residence wasn't good, was dangerous, and because

---

3. Romual'd A. Pogaevskii was a Sakhalin physician who, at the time of Akif'ëv's visit, also led the island's Medical Division.

4. This may be a reference to the Moravitskii Case, which involved corruption charges against a former Sakhalin prison warden, Aleksandr E. Moravitskii, who was eventually found innocent.

we have several thousand rubles with us, we brought them to the treasury this morning. As a matter of fact, the morals here are utterly savage: since we arrived there have already been three murders. A wife who had voluntarily followed her exile-settler husband from Russia had found herself a lover, and she and the lover finished off the husband and buried him under a shed in the yard. The prison director Patrin is well versed in the Sakhalinians' tricks and immediately found them out.[5] Coming into the home, he noticed a spade covered with some fresh dirt in the entryway. He likewise immediately inspected the whole yard, and going into the shed, saw that its floor had been dug up and only recently tamped back down; he ordered it dug up, and the corpse of the murdered man proved to be there, folded in half. Then the bodies of two watchmen were found in one of the ravines near the same town.

In general, the murder of someone here has apparently lost its terrible impact, its horror. Rarely does a day pass without a murder.[6] They kill for a tuppence, for a heavy pea jacket, and so easily, God knows why. Life has evidently lost all meaning; it is absolutely absurd. Imagine a man sentenced for life to penal labor: he lives in a filthy prison one, two years, suffering beatings, birch-rods, lives in a society of similarly unfortunate and embittered people, knowing he will not be released, and because he wants to breathe free air, he yearns to leave the prison and senselessly escapes, runs knowing that if a watchman sees him he will shoot him; he escapes knowing that Giliaks[7] and guards are hunting him, knowing many fugitives die from exposure in the Tatar Strait or, even farther in the mountains and forests of Siberia, from wild beasts, frosts, and starvation. Despite all this, the man decides that if he cannot stop in the face of a thousand dangers, if he does not value his own life, if, in other words, freedom is for him dearer than life, that he will stop only if another man crosses his path, and then only perhaps. Escapees are captured, flogged with whips, put in chains, and given extended penal labor terms, yet they continue to escape. The environs around Aleksandrovsk are full of escapees, and they commit murders.

This afternoon around two o'clock we visited the place of M., director of the governor's chancery.[8] It happened to be on the occasion of a philanthropic concert in which anyone capable of anything participated. There I met the jurymen of the military district court. The military court has come here for the case of a murder of three convoy guards by five escaped penal laborers. The murderers were captured and are in solitary confinement. It's said they're anticipating a sentence of death by hanging. The phrase "death sentence"

5. Aleksandr Ia. Patrin was renowned for his harshness.
6. Sakhalin's murder rate was indeed high, but this is a gross exaggeration.
7. Giliaks are native to Sakhalin and the Amur region. Giliaks are today called Nivkhi.
8. This was a man name Marchenko.

horrified us, but they're so habituated to it here that they hold a concert in the same courtroom where death sentences are issued, and they perform on the very same stage from which the horrible sentence is pronounced.

What a staggering practice![9]

We happened to hear about local villainy once again. Robbery and theft exist to a high degree here: storerooms' contents get cleaned out annually, sometimes several times. But there's something more brazen. Not long ago several people armed with revolvers were in one of the shops, practically in the center of town, and were demanding money; then they ransacked the shop and took what was of value. This happened in broad daylight. After someone alerted the police, they ran to this shop, having grabbed their rifles, but the robbers seemed to have disappeared into the ground. Some exile-settlers obviously helped them in this. There are some stunning frauds here: during winter an official somehow purchased a ram, which is very rare here, and took it home; what was his surprise when this ram proved to be nothing other than a dog to which the hoarfrost had given a ram's head.

Our Miagkov has already been suckered. Some character offered to make a suitcase out of seal hide; they agreed on 14 rubles. He then took the measurements and asked for money to buy the materials. Gennad'ich gave him half, i.e., 7 rubles, and asked that it be made quickly. The character vowed and swore he'd get to work right away and that everything would be alright. That evening, he appeared before me already quite drunk and with a seal hide, and showing it to me, asked whether it was alright. I praised the hide and ordered him to go home and get to work. Suddenly he said:

"Please give me another two rubles."

"What for?" I asked.

"Um, I don't have enough materials"

"Well, brother, you've been given enough money, go and sew."

He rolled up the hide and left.

I hadn't seen him for a long time when he walked past the porch and laid the hide out on the pavement. It turned out he'd managed to persuade our senior laborer Petrov to give him two rubles. He'd gone on a binge, and of course never gave Gennad'ich any kind of suitcase.

Today we were at the home of Doctor Volkenshtein, the local hospital's psychiatrist, and became acquainted with his wife, an unfortunate woman who

---

9. As with Sakhalin's murder rate, Akif'ev implicitly exaggerates the number of executions there. Between the mid-eighteenth and early twentieth centuries, the number of legal executions in Russia was low compared to most other European countries, though this changed after 1905.

spent 14 years in solitary confinement and was then exiled to settlement here. It's simply amazing how much this person has endured![10]

Two more politicals are here: Khronovskii and Manucharov.[11] The former is now in the free estate but hasn't left for the mainland, he serves in the capacity of schoolteacher here and moreover writes about Sakhalin for one of the Vladivostok newspapers. The other is still an exile-settler and serves here as editor for the local weekly organ *Russian Agency Telegrams*.[12] In general, educated people[13] are highly valued here, and educated penal laborers and exile-settlers are not forced to do common labor but are given minor chancery and other jobs where they can be more useful.

In the evening, we were again at Engineer K.'s place and heard in abundance about life on Sakhalin. A penal laborer's entire life, his relationship to the administration and vice versa, is a sort of *circulus vitiosis* [vicious circle].

20 September/3 October. Last night the concert was held in the courthouse, where there was playing, singing, and declaiming. The concert made a strange

---

10. Aleksandr A. Volkenshtein met and married Liudmila A. Aleksandrova during the early 1870s. While enrolled at Kiev University, he was arrested for participating in the Going to the People movement during the summer of 1874. He was tried but acquitted during the famous Trial of the 193 (October 1877 to January 1878). The couple's relationship soon soured, and Aleksandr left to serve in the Red Cross in Bulgaria during the Russo-Turkish War. Left alone with their young son, Liudmila became increasingly radicalized and joined a conspiracy that led to the assassination of Kiev's governor-general in 1879. Aleksandr returned home soon thereafter. Liudmila, aware that she was wanted by the police, left her son and husband and fled to Europe to avoid arrest. But three years later, when she returned to Russia, she was arrested and sentenced to death—a sentence immediately commuted to fifteen years' solitary confinement. She would eventually serve twelve years in the Shlissel'burg and Petropavlovsk fortresses. Aleksandr remarried, worked as a *zemskii* doctor, and joined the Tolstoyans. One day he received a letter from his ex-wife, telling him she was being released three years early but would be exiled to Sakhalin. Unhappy with his new wife, Aleksandr obtained permission from St. Petersburg that allowed him to follow Liudmila to Sakhalin a couple years later. Lacking qualified personnel, the island administration immediately hired him as a doctor. However, other sources indicate that instead of being a psychiatrist, Volkenshtein worked as a general practitioner for the convicts. Liudmila herself worked as a paramedic and meteorological observer, and later as a journalist. In September 1902, the couple received permission to transfer to Vladivostok, where they were joined by Aleksandr's wife and son. On 10 January 1906, Liudmila joined a public demonstration to mark the one-year anniversary of Bloody Sunday. Soldiers opened fire on the crowd, and she along with twenty-nine others were killed. See M. V. Teplinskii, "Doktor A. A. Volkenshtein (shtrikhi k portretu obyknovennogo poriadochnogo cheloveka)," *Vestnik Sakhalinskogo muzeia* 4 (1997): 145–87.

11. "*Politicheskie*"—synecdoche for "political exiles." Stanislav F. Khronovskii arrived on Sakhalin in 1888. Despite becoming eligible for transfer to the mainland, he had chosen to remain on the island, which is why Akif'ëv was able to meet him there. Ivan L. Manucharov arrived on Sakhalin in 1896, having like Liudmila Volkenshtein earlier been in Shlissel'burg. Akif'ëv was apparently unaware there were actually around a dozen political exiles on Sakhalin at the time of his visit.

12. This was really just a digest that the official print shop produced from the telegrams sent to Sakhalin.

13. Educated people (*intelligentnye liudi*)—that is, the Russian intelligentsia.

impression: a day earlier, there was a trial there; and tomorrow the familiar sentence: "Such a many lashings, and then the death penalty" will be given; but now the hall was full of people. Administrators, engineers, and officials, as well as penal laborers and exile-settlers, were all sitting alike on a row of stools and simply forming an audience. Who the performers were was completely determined by the circumstances: the chancery director declaimed to the accompaniment of a convict's wife, the assistant prison warden played the balalaika to the chancery director's accompaniment, the military court's defense counsel played the violoncello, a tourist who had turned up sang, and the orchestra of soldiers and exile-settlers droned on. In a word, various elements coalesced within the realm of art. The circumstances were unusual, the performers and audience unusual. Strangely, a rather terrible but nevertheless pleasant understanding managed to unite these poles, and though they would, within several minutes, separate to different positions, there were for now no wretched penal laborers and exile-settlers or ominous jurymen and strict administrators, just simply performers and an audience.

Following the concert some of the audience and performers repaired to the club to dine; but no one else wanted to break the mood, so it was decided to continue with winnowed singing and music playing. Guitars were sent for and a choir began singing Gypsy songs. The singing wasn't harmonious, but it was cheerful and heartfelt; Gennad'ich and I also sang and even performed a duet. One official was sitting and listening to us attentively, and then went over to comrade prosecutor Semënov and enigmatically whispered in his ear:

"These visitors sing quite well, so it wouldn't be a silly thing to inspect their passports to learn who they really are."

He couldn't restrain himself and burst out laughing, and the entire company, having noticed this business, teamed up against him, and in this way the evening came to a cheerful close.

Dr. Volkenshtein came this morning and we went to see the prison. It should be said we received permission for this last night. When we reached the prison, F., the assistant prison warden, noticed and came out so as to escort us. We were first of all led to the so-called chains prison. This prison is for probationers and those who frequently escape and who, so to speak, are unreformable.[14] The prison presented itself as a rather large square guarded by a tall stockade, gray from age and dilapidated in spots and behind which nothing could be seen. With a sinking feeling, we approached the gate, behind which we knew was a crowd

---

14. Russian penal labor (*katorga*) consisted of two prisoner categories: the probationers (*ispytuemye*) were held under the most severe restrictions; following several years of good behavior, they graduated to become correctionals (*ispravliaiushchiksia*), who enjoyed fewer restrictions and greater privileges.

of people deprived of all rights to a human way of life. It became really terrifying. Then the gates opened with a creak and we entered a large, wide yard.

Low and quite decrepit, the penal laborers' complex had been constructed along a wall, deviating from it a bit. Penal laborers in the most varied, sometimes tattered, uniforms, with heads half shaven, were strolling back and forth along the spacious yard, and fetters were ringing and clanging here and there. A guard appeared alongside us and shouted "Attention!" at the top of his lungs and the inmates removed their caps. All these people standing before us without their caps were extremely pitiful. They had once been people, many with a house and family, but now there was none of that, and no freedom, and their legs were chained in fetters and their heads disgracefully half-shaven, and every guard—often an exile as well—could flog them like Sidorov's goat,[15] and this they had to endure, the future a darkness without light. A man endures penal labor, graduates to a settlement, completes his term, and becomes a former convict. Even if he completely morally reforms himself, this stigma remains for his entire life and he will never be completely equal.

We walked through the complex. All the buildings were old and dark and there were many people. In one section, where sat the so-called *promotchiki*,[16] 25 men were packed into a space comfortable for no more than 12. I could imagine what the air was like there during winter! Even with the doors open, the air was thick. From there, we entered the cell where the convoy guards' murderers (who've been mentioned) were. They knew they were awaiting execution, and this was evident in the utterly murderous face of one of them summoned from the cell. He spoke sluggishly and quietly, in a shaky voice. But the other confederates were lying on their sleeping platforms in the dark cell, illuminated by a single, small, grilled aperture opening to the corridor. They were lying facedown but obviously weren't asleep, since they were surrounded by the tread and talk of the guards. It was horrible to see them and to think that in several days, they will be swinging from the noose. To crown the horror, F. laughingly pointed at them and barked: "Ah, here are the candidates." These words cut me like a knife, and I hurried to leave the cell.

We went to the next barracks and the so-called wheelbarrow-men were summoned. These criminals have escaped and killed several times. They're attached to wheelbarrows by a long chain extending to the waist and, in this way, they must push their wheelbarrows wherever they go. Lying down to sleep, they place the wheelbarrows beneath the sleeping platform. Besides this, they wear

---

15. The colloquial phrase "like Sidorov's goat" means "like a whipping horse."
16. *Promotchiki* is what those penal laborers who have lost government clothing through card games and dice and who generally squander it away are called. [Akif'ëv's note.]

manacles. In such way they spend several years with a wheelbarrow. One elder, 56 years old, was brought out, and he was difficult to look at. Were there really not enough shackles on him? After him another, who not long ago was freed from his wheelbarrow, made a strong impression. This was a tall, solid man of middle age with a repulsive visage: his hair and beard were red, his lower jawbone stuck out, and his forehead sloped backwards; his small gray eyes looked at us coldly and hostilely. Hence, it was clear that were you to be alone with him, he wouldn't think twice about grabbing your throat. His name was Shirokolobov; he's been chained to a wheelbarrow for nine years. He's escaped many times and, it's said, is responsible for thirty murders and maybe more. He was loathsome to look at.[17]

We were passing through the yard. A penal laborer approached us, took off his cap, and beseeched Volkenshtein: "Master, take me to the village, I'm bored here!"

"But what will you do in the village?"

"I'll plow and harrow, Master, take me with you, for Christ's sake!" he tearfully said at the top of his voice.

"Very well, very well, my dear, I'll ask," and Volkenshtein walked on.

"Why, he's mentally ill!" I said.

"Yes, unfortunately there are very many here like him. Look around, you'll see."

We entered a building. Eight people were sitting on a sleeping platform in a small room. This section was for the nonviolent ones, whereas the violent ones were housed in the psychiatric hospital. It was difficult and sad to see these wretches: they could not account for their situation or why they were there at all. And perhaps they'd not been in a fully normal condition when they committed their crimes. If so, sadder still!

A lot of penal laborers were walking around the yard.

"Why is it they don't work but just wander from corner to corner?" I asked.

"Many refuse to work, but the rest have been released from it for some reason. In general, we don't especially trouble ourselves with work," answered F.

The chancery director Mr. M. had told me something similar last night.

"Strictly speaking, is there even penal labor here? They don't bother to work, and the sick are always freed from labor over trifles."

On the other hand, I heard this contrary opinion:

---

17. Fëdor Shirokolobov was indeed one of Sakhalin's most renowned prisoners. Stories about him vary but universally agree he was a serial killer. Akif'ëv's description of his features and the conclusions he draws from these suggest the influence on him of Lombrosianism, which held that criminal "types" could be discerned by their physical characteristics.

"There's a lot that's gotten out of hand here, and the leadership can't do anything with us other than put us in chains without work, so the rest of us gotta do a lotta work; in sum, the work's not being evenly distributed here."

We went from the chains prison to the chains infirmary. This was a low, quite dark building, but nonetheless better than the complex. To our questions as to how things were and how often the doctor visited them, the patients responded that he visited every day and they generally spoke about him with respect.

From there, we exited the chains and strolled along the road to the place where the hangings are to be carried out. The spot chosen for this is a passage between a fence and one of the barracks, right in front of its windows.

"Why did they choose such a spot, visible from the windows right there," I said, "because, really, a capital execution doesn't seem to me to constitute a moral corrective."

"It's just customary, but to make it less visible we conduct executions at dawn, while the penal laborers are still sleeping. The gallows are built at night and quickly dismantled after the execution," replied F.

Having finished seeing the chains prison, we went to the so-called "free" prison. This rather sarcastic name signifies that the correctionals are placed in this prison until the conclusion of their terms and enjoy relative freedom. They can labor on the outside and need only be present for morning and evening roll calls. At that moment it was deserted, everyone was at work, and nowhere was there the sound of chains. Walking around some barracks that were—it must be said—generally utterly rotten, we came to the almshouse. The sight was terribly depressing. Encountered among some still quite young people were those who struck you as being just ghosts. One ancient old man could not rise upon our entering, and so his hunched-up, decrepit, emaciated body continued to sit. He didn't even notice us. And there was a man groveling on his knees on the floor. This wretch's legs had been frostbitten during an escape and had had to be amputated. Curiosities were to be encountered here. I asked: "How old are you?"

"Well, I'd be less than sixty."

"And how many years left in your sentence?"

"Another 56 years."

Not bad. Sometimes, you'd hear such an answer as follows:

"I got life, plus another 20 years."

At first, you couldn't tell if they were joking. Honestly, they proved not to be. It's said there's even one case where a certain penal laborer accumulated 115 years of penal labor. A penal labor term usually increases because of escapes. There are stories such as the following: a man gets sentenced to 5 years' penal labor and sits for several years but can't restrain himself—he flees. He's captured, flogged, and given several additional years of penal labor, and he flees again, and again he's captured, etc. In a word, dozens of years later, he's practically forgotten the crime he was exiled for, but continues to sit in irons for wanting his freedom. How many people must remain here until death, and what must their lives be if they don't have the slimmest hope of something better in the future? It's truly repulsive and horrifying!

That people escape is not at all surprising. A very small portion crosses the Tatar Strait and hides on the mainland. Most are captured and put in prison again or are killed. A hunt for fugitives actually exists here. Giliaks and soldiers receive three rubles for each fugitive captured or killed. Giliaks, those denizens of the forest, terrify fugitives by the fact that they can track them through the woods and steal upon them completely unawares and unnoticed. Soldiers, resenting escapees for murdering their comrades, sometimes shoot them down as a group. There was recently an instance when soldiers avenging the deaths of three comrades slaughtered 17 fugitives.

Here's what happened. A horde of fugitives came to a Giliak village, and threatening the Giliaks with death, demanded to be brought across the Tatar Strait in boats. The Giliaks seemed to be complying, but they instantly dispatched a secret messenger to a guard post 60 versts away and took as much time as they could to prepare the boats for crossing. Finally, everything was ready and they went to sea. The penal laborers were alarmed that they weren't going directly across the strait but were hovering near shore, but because they weren't experienced in maritime affairs, the Giliaks managed to convince them the wind was such as to make sailing impossible. Meanwhile, soldiers from the post who had been alerted by the Giliak had already reached the shore and positioned themselves behind a cliff. When the boats were level with the ambuscade the Giliaks, on a predetermined signal, dove into the water and swam to shore, but the soldiers killed all the penal laborers, firing as if they were targets. This well illustrates Sakhalin morals.

I return to my diary. We went from the almshouse to the mess and then to ward no. 10, where corporal punishment is administered. This was divided into cells whose doors led to a wide corridor. Standing in this corridor was the mare [*kobylka*[18]], that is, a wide bench with notches for the arms and legs, somewhat inclined, on which the victim is laid down and tied. Here we met the executioner, a tall, lean man, and we were at the same time shown the birch rods, lash, and the straps for tying down. One sight of these instruments caused a shiver;

---

18. Also translatable as "filly," this was prisoner jargon endowing an instrument of torture with a sexual connotation. L. V. Belovinskii, *Rossiiskii istoriko-bytovoi slovar'* (Moscow: Studiia "TRITE" Nikity Mikhalkova, "Rossiiskii arkhiv," 1999), 204; and Andrew A. Gentes, "'Beat the Devil!': Prison Society and Anarchy in Tsarist Siberia," *Ab Imperio* 2 (2009): 201–24.

it's hard to say what would be worse, the rods or the lash. A rod is four wooden switches half a finger's width each, twisted together. Since they are for the most part knotty, it's no wonder they cut the skin with the first blow. The lash is rather heavy and has three two-*arshiny*-long straps. Each ends in a thick knot. Regarding this terrible instrument, the executioner expressed himself as follows:

"A man can be beaten to death with five blows of the lash."

"But can *you* kill with five blows?"

"No. What am I? But Komlëv could. Still, I can break a spine. But Komlëv was sharp and quick, one blow after another, and he didn't give them a chance to breathe; well, he had the spirit."[19]

"But how many lashes have you yourself received?" Volkenshtein asked him.

"I've gotten five hundred, yes, and from Komlëv."

"Well, but do you flog often?"

"It's rarer now," F. interrupted. "The doctors interfere now, but we need to make them toe the line."

"But are they really *interfering*?"

"Yes, of course, or they release them entirely, saying they're sick, or they limit the number of lashes."

"Well, but do you often strike hard?" we asked the executioner.

"No, it's much weaker now according to the new regulation. Now, your elbow can't leave your body, but earlier, you could swing from the shoulder."

And he gave a demonstration of how they used to strike and how they strike now.

"Yes, thank God the doctors have begun freeing them," he added.

"Well, tell us plainly: is it unpleasant and difficult for you to flog or not?"

"How can't it be difficult, Your Worship? I'm a man and I was flogged too, so I know it ain't sweet. But there's other times, when you're bitter, and you strike hard. Here it is—either you beat your comrade or some other swine's gonna do it."

"No pity for these scoundrels," F. interrupted him, "if'n they won't pity their own. Should be fewer swine. But while they was gambling and singing one evening, three watchmen got bumped off in that ravine over there."[20]

All these murders, executions, lashes, birch rods, and clanging chains were so unusual and horrifying to us that our gift of speech vanished, and we could only look and listen. And listening to the executioner and Mr. F. there, I unconsciously

---

19. Nikita Komlëv was Sakhalin's most infamous executioner (*palach'*).

20. On Sakhalin, exile-settlers typically served as watchmen. The story F. tells is partially confirmed by a newspaper article from around this time, which, however, reports that *two* watchmen's corpses were in the ravine.

compared their words and realized, to my horror, that the executioner seemed more humane than the assistant prison warden, a man with authority over some two thousand prisoners. What influence does he have on them? Surprisingly little, I would suppose. He's risen from the lower ranks and gone straight from the soldiery to being assistant prison warden, and he feels himself powerful, imagines himself important, and considers himself something special. It's sad that such people are assigned to such positions, though contrarily, few educated and proper individuals can be found who will take such a position.

From the "free" prison we went to the women's prison. This was a small, standalone building, with a separate entrance off the square and a small yard attached to it. It had space for 25–30 people and was cleaner than the men's prison, with separate bunks, though only 6 to 8 inmates were there.

"Why so few here?"

"They've all been distributed as either domestics or cohabitants and only these here are left."

One was a young woman, 20 years old and strikingly beautiful, with a characteristic southerner's face, but with such a gloomy expression in her large eyes, burning as if with fire, that I understood why no one had taken her into cohabitation. There was also a young mother with a year-old baby in her arms. Gennad'ich approached and began playing with the baby; the baby tugged at his hands and laughed, and its mother laughed. Suddenly, F. quietly told me: "But this one here, with the baby, is the same one who with her lover murdered her husband 5 days ago." I was simply horrorstruck. How was this possible? How could she, only five days ago, have killed her husband, whom she followed from Russia to Sakhalin, the father of her baby, and now be here laughing, laughing easily and joyfully? I could take no more and hurried to leave the prison.

This women's prison added to the bitter cup of the day's impressions. There is such a terrible incongruity. A woman commits a crime, a court sentences her to punishment, and she is exiled to Sakhalin. A picture of the distribution of female prisoners who arrive here can be sketched as follows: a steamer arrives, the women are brought to Aleksandrovsk Post; the top people and administrators instantly appear and choose for their service and according to taste; after them come the guards, and they, too, choose their cohabitants according to taste; then the exile-settlers and penal laborers arrive. The inspections and conversations are swiftly concluded. The leftover women are sent to the other districts for cohabitation; but if they remain unstationed, so to speak, they are incarcerated in Aleksandrovsk Prison. No suitable or interesting work for women has yet been devised here, and so rather than die from boredom in prison, the women prefer to enter cohabitation and at least some measure of freedom. In such way, force of circumstance compels a woman to become a cohabitant to someone she has seen once and may not love. What does this lead to? A woman ceases to respect herself and this leads her to depravity. And it is widespread here like nowhere else: this filthy, cynical depravity. This is penal labor, the goal of which should be to rehabilitate, but on the contrary, it promotes depravity and ruins a woman.

All of Sakhalin is infected with a well-known disease.[21] This is all reflected in how the children are: you see these wretched, pale, skinny children, you hear what they say, and your hair stands on end. A seven-year-old boy is smoking a cigarette and swearing with the choicest words, completely unashamedly. And here's a group playing fugitives and soldiers, playing at robbers, executioners, and prison guards and wardens. What cute games these are, no?

21 September/4 October. Today around noon, Gennad'ich and I went with the engineer Kalistov to see the coal mines. Doctor Volkenshtein joined us, and so the entire procession consisted of three carriages. The distance wasn't far, all of one and a half versts, so we alighted from the carriages and quickly walked to the galleries. According to Kalistov, there's a whole bunch of coal here but it's often mixed with schist. Little of it is being processed, and mostly only for Aleksandrovsk Post's needs, though strictly speaking, the supply could be increased. Vladivostok is not far away, and the demand for coal there is large and always will be. But the matter of production was earlier quite poorly conducted, and now the works operate as if to show they can function nevertheless. Engineer Kalistov, who has recently arrived here, began with what should have been started long ago, that is, basic research and a study of the wide-ranging coal strata. The terrain here is very broken and the strata contorted. Their angles are sometimes very steep, approaching nearly 90°. The work was being done without plan or planning, and this is why Kalistov's work was appreciated with delight.[22]

The business very much interests him: he's already knocked out a bunch of prospecting shafts, has developed for himself a schematic of the coal strata's paths, and is developing a plan for extracting the coal. For several hours, we followed him with interest, climbing the mountain and passing through the galleries, he explaining the whole time how the work is done and what his research has yielded. We'd come at lunchtime and so were unable to observe the gallery labor. We were given candles and went deeper into the mountain.

---

21. After our Sakhalin visit a quarter of our laborers appealed to me for advice. It turned out they were all infected. [Akif'ëv's note almost certainly refers to syphilis, which was indeed widespread on Sakhalin.]

22. Prior to the 1890s, a joint-stock company called "Sakhalin" held monopoly mining rights. It was poorly managed and spectacularly corrupt.

The galleries proved to be very low and narrow corridors: we had to bow our heads and practically crawl on the floor in places. Imagining that people spend their workdays here in the damp and darkness deep below the earth made for an unpleasantly heavy impression. A lot of moisture was flowing down the walls and puddling on the floor in some galleries, though the air was not particularly oppressive and the candles burned bright. This was because good ventilation had been constructed there. But in one shaft where there was no ventilation the air was close and heavy, the candles gave only a weak light, and breathing was difficult.

In one of the shafts, the coal is loaded in wagons and pulled along rails, but in the rest, it is upon extraction loaded by hand into wheelbarrows that are remarkably heavy and awkward. Exiting the final shaft, we were approached by fettered parties with wheelbarrows for the coal, which they delivered to the wharf. There was one barrow for every 8 laborers, who pulled it along the beaten-down spots and the dirt road.

It then became clear what penal labor is: this is hard work accompanied by a huge loss of time and effort, and it is completely inefficient because it is so stupidly designed, and the time and effort wasted could be less and the results achieved much greater. The laborers themselves understand this perfectly, so the work not only dulls their interest but becomes boring and offensive to them. Indeed, given the arrangements, 8 laborers pull a heavy barrow of coal along a bad road. The distance is all of 1 1/2 versts, i.e., not far at all. But they manage to bring only 3 barrows to the wharf in the course of a day, and during this time they take breaks. Yet since the area is completely level, rails might be laid from the mine to the wharf, because the timber for the railroad ties is right there and there's no shortage of labor, and if no technical difficulties were to be encountered, the rails just need to be ordered and construction could begin. It would go like this: whereas 8 people now accompany a barrow, only two would be needed for such a barrow, even the heaviest, and if they accompany it, not three times, but 6 or 7, and were they to exchange pulling it by hand for a horse, the business would go still faster. This would free up many hands, which would increase the extracting of coal; and then a laborer might feel he's working intelligently, that he's performing an essential function, and the work would stop being boring and offensive; revenues would increase, such that laborers' lives might be turned around and improved. And what a laborer's life needs for improvement we witnessed soon enough.

Having inspected all the mines that could be looked at, we repaired to the barracks in which were housed the laborers from the "free" prison. Kalistov warned us we would encounter something very bad, but the reality proved far worse than horrible. The barracks were nothing other than small, dark izbas, half submerged in the ground. Their exteriors were very bad: windows broken, walls buckling, roofs ruined, yet the impression we received upon entering these barracks cannot be put into words. Once inside one, we were struck by an especially sour sort of smell. There was a stove in one corner, really not a stove but just a pile of bricks on which something is seared or boiled; the smoke and fumes went straight into the izba; the walls were simply astonishing, more like a lattice than a wall; the chinks that had formed between the timbers were up to two *vershki* wide in places. I assure you I am not exaggerating. Contiguous sleeping platforms ran down the middle of the izbas; there were no floors, just boards laid on the ground that clattered and moved like the keys on a pianoforte. These izbas were filled with people. It was hard to imagine anything worse. People come home from dark, stuffy shafts to rest and satiate their hunger, and they fall, not into an izba, but rather this sort of pit that condemns everyone to communicable diseases.

"Gentlemen, have these abodes gotten your attention? Imagine all the horrors of living in these barracks during winter. What the temperature and the air would be like in here?" said Kalistov.

"Your Worships, please look at the bread they're giving us," a laborer holding a piece of rye bread appealed to us.

The bread was completely soggy, underbaked, and as sticky and tacky as putty.

"Your Worships, please tell them to thoroughly cook the bread, otherwise our stomachs get sick from this soggy stuff," requested the laborers.

Having promised to speak about this with someone, we left the barracks feeling terribly despondent.

Judging from the prison, the operations, and these barracks, can it be any clearer why, properly speaking, Sakhalin's operations have gotten badly out of hand and why penal laborers are so demoralized? Really, do they want penal labor to rehabilitate and not utterly ruin people, do they want these people to be useful, so they can be appealed to as people, and to not destroy the last feeling of human dignity in them? And if half-starved people are held in kennels, fed God-knows-what and what cannot be called decent bread, and if they are viewed with supreme contempt, like beasts, then what good can be expected?

Penal labor ruins a person. In court after court in Russia I've happened to see criminals, and most either repented of their crimes or were terribly shocked by their sentences: cynicism and braggadocio are rarely encountered. Here, those who get the most lashes, who've escaped the most often, who suffer the most, do the boasting; the air is filled with the words: murder, robbery, lashes, birch rods, prison, escape, hangings; and whereas your—the novice's—hair stands on end, the local residents, penal laborers, and administrators are completely inured to it. You get brutalized and numbed. One day, in a moment of candor, Mr. F. said: "You know, I haven't lived here long, but I feel like I'm starting to get brutalized, yet my commander has been completely brutalized." There was this on the one hand, but on the other, his attitude toward the penal laborer is like that toward a basc, powerless creature, and his cognizance of this powerlessness, this grim and hopeless cognizance, destroys in him his better spiritual powers, and he truly is becoming brutalized.

23 September/6 October. Yesterday morning the steamer *Sungari*, on which we're to go to Vladivostok, arrived. Having made our farewell visits, Gennad'ich and I went with our laborers to the wharf. The weather accompanying us was as severe as has been encountered: the wind wasn't especially strong, but the chop was powerful and we climbed with difficulty the jacob's ladder to the steamer, though the comrade prosecutor, who has an amputated arm, had to have a rope lashed round his torso and was, in this way, lifted onto the gunwale. Dr. Volkenshtein accompanied us and helped load our baggage. During our brief acquaintance he impressed us as a very kind and intelligent man, and we shook his hand warmly, having wished him as speedy a leave as possible from this terrible island. Mr. F. also came, and he caught one of the penal laborers who was loading coal using a counterfeit coin.

Here's what happened. A penal laborer was speaking quietly with a Chinese buffet attendant and bought a bottle of liquor for 5 rubles, for which he gave him a gold five-piece. Only afterward did the Chinese notice the coin was somehow very light and the eagle not clearly demarcated. He then turned to Mr. F., who quickly found the villain, who immediately confessed.

Dr. Salmanov, a senior military physician, and Dr. Bulygin, my friend from university, were to come see me off, but the chop had so picked up that the cutter could not risk mooring to the steamer, and it was also very dark, so they turned back. Today, it turned out we wouldn't be leaving until 8 o'clock in the evening, so Gennad'ich and comrade prosecutor Semënov left for shore. Salmanov and Bulygin came to the steamer around noon.[23]

How nice it was to see this man I'd not seen since university, and to be greeting each other 10,180 versts from Petersburg as if we were at a telegraph pole outside the Aleksandrovsk post office. It turns out he's lived in Dué[24] for two years, in the capacity of military physician, and has another one and a half

---

23. Pëtr P. Salmanov (b. 1865) led Aleksandrovsk Post's military medical unit. Sergei E. Bulygin (b. 1871) was a military physician assigned to Dué Post.

24. Located a few miles south of Aleksandrovsk Post, Dué Post was Russia's first settlement on Sakhalin, as well as the location of the "Sakhalin" Company's mines and the earliest Russian prisons there.

years to go. I don't envy him. Our conversations and reminiscences lasted all day, and as the steamer was getting ready to leave our tourists[25] still hadn't shown up, so the captain began blowing the whistle to summon them. They finally returned around 8 o'clock and the steamer, anchor raised, left for Korsakovsk Post.

This evening we passengers got acquainted. Traveling with us is a session of the Vladivostok Circuit Court, a forest inspector and his lively wife and daughter, and a just-married officer now going to the war in China. We got to know each other and had a cheerful evening of singing and talking about our group of shared acquaintances, grateful we aren't on Sakhalin but at sea and not witnessing all of Sakhalin's horrors, and we then dispersed to our cabins.

25 September/8 October. Korsakovsk Post. We arrived at Korsakovsk Post this morning; technically, we were supposed to stop at Mauka (a seafood enterprise on Sakhalin's west coast[26]) but couldn't because of a strong wind, so we came straight here. It was a beautiful day: the winds had calmed, the sky cleared, and Gennad'ich and I decided to go ashore.

Korsakovsk Post is situated in a deep cove on a hilly shoreline. A narrow street with planted trees runs from the hills to the sea; neatly constructed houses disappear amid the greenery. In general, the scene is very pleasant, and it would be impossible to suppose there is a prison here no less terrible than in Aleksandrovsk. A wharf has been built on the shore and it is complemented by the ships being kept there, as well as two cutters.

We crossed to the shore and began wandering the streets. Entering one shop to purchase cigarettes, we struck up a conversation with the proprietor and he suggested we buy three miniatures of penal labor life that a penal laborer here creates out of painted bread. This collection was splendidly done: in the figurines of penal laborers and soldiers, their movements and all details of their uniforms and even facial expressions were wonderfully portrayed. Two of the collection show penal laborers doing wintertime work: one is of a penal laborer in patched pea-jacket and trousers dragging a log to portage, the other one is a penal laborer driving a sledge loaded with firewood. They are as solidly made as they are detailed. The third miniature of the collection portrays a punishment with birch rods: two soldiers are holding the condemned penal laborer down on the mare; his body is covered in bloody stripes and blood is running onto the mare. Above him stands the executioner, getting ready to strike with a fasces of

---

25. "Tourists" is a joking reference to Semënov and Gennad'ich.

26. This was Semënov and Co., formed through a partnership between a Russian and a Scot named Denbigh, which sold large volumes of sea kale and fish products, principally to Japan.

birch rods in his hand, splintered rods are spread on the ground, and there stands beside him a fat, well-fed gentleman in the form of a prison commander. So realistic are the poses and coloration that it seems the tortured man will start screaming. We acquired this collection immediately.

We left the shop for the steamer and arrived for lunch. As soon as we sat down to eat there appeared a rather young officer, in a new, clean dress uniform, smoothly shaven, pink (simply blood with milk), with a considerable belly and something Olympian to his face.

"Who's this?" I quietly asked.

"The local prison commander, Mr. Sh.," they answered.[27]

He sat beside the captain and immediately began discoursing about—of course—penal labor, punishments, and so on. Sakhalinians seem unable to go ten minutes without launching into their favorite topic. This, so to speak, is the crux of their madness. I listened to the conversation.

The prison commander dashingly knocked back a tumbler of vodka, took a bite to whet his appetite, and continued.

"Yes, penal laborers are undisciplined, terribly disorganized, and much work will be needed to bring everything into order; but I'm hopeful, nonetheless. I oppose corporal punishment; it degrades and ruins a man, and I prefer solitary confinement; take, for example, this young man Gorskii I have—he doesn't want to obey and, moreover, bothers all the prisoners. I put him in the isolator and will keep him there until he submits; I break them, and if he doesn't submit within a month I'll hold him for two or three or, ultimately, a year or two, but all the same, I break them."

"But do you really have the authority to hold them in solitary confinement for a year or two without the court?"

"Of course I do, and no one has the right to interfere."

"Strictly speaking, as far as I know, shouldn't the prosecutorial inspectorate supervise the prison commander?" P., a member of the court, quietly remarked from the other end of the table.

"Of course, of course," Sh. suddenly recovered. "I wanted to say that I've been given the *right* to order someone put in solitary confinement . . ."—and his peroration forcefully gushed anew, and he talked so much that Gennad'ich, who'd been seated next to him, could no longer restrain himself and stood up from the table. At that moment, Semënov arrived and suggested we go ashore and, if desired, visit the prison he was now going to walk around. The prison commander very courteously invited us as well. We left.

---

27. Viktor V. Shel'king (1847–?).

Onshore, we began by visiting the military physician D.[28]; we sat awhile in his place and left for the prison in his company. We'd hardly reached the prison gates when a curious thing happened. The comrade prosecutor ordered the soldier on watch to open the prison gates.

"No one's allowed to pass," came the answer.

Semënov actually stepped backwards out of perplexity.

"*Who* has forbidden entry?"

"The prison's gentleman commander."

"But you should know I'm the comrade prosecutor and I have the right to enter."

"I don't know, no one's ordered me to open up."

In a word, the comrade prosecutor wasn't being allowed into the prison. Sh.'s quarters were just opposite the gate. Semënov asked us to wait and went to them. After several minutes they both came down the steps.

"I didn't order any such thing," said Sh. "This fool confuses everything. I only ordered him to report to me who is trying to visit the prison."

In other words, he didn't want anyone in the prison without him.

"Well, gentlemen, please!" The gates opened with a creak, and we entered the yard. The yard was clean and had buildings along its sides: there were a barracks, an almshouse, and a joinery shop. Across the yard stood some kind of old building, and just behind it was the chains prison. Going over there, we entered the primary structure.

This was a building for those under investigation; it was quite new, tall, and sufficiently roomy, though not especially bright, but generally incomparably better than what was in Aleksandrovsk. In the first hall we found 25 inmates, some in chains, some not. They all stood upon our appearance and, seeing Semënov, immediately circled him and began asking and pleading about things. The prison commander, puffing a cigarette and tossing back his head, peered down at the inmates; but at the same time, he didn't bother listening to what they were telling the comrade prosecutor.[29] This chamber held primarily escapees.

"Gentlemen, focus your attention on these rascals: they escape from one prison to another. Orderliness is observed here, if you please . . ."—said Sh., pointing out the prisoners to us.

---

28. Nikolai A. Davydov, a military psychiatrist who, despite his posting, also treated prisoners, among whom he had a reputation as a sadist.

29. Anton Chekhov describes Shel'king as "a large, portly man, with that solid, imposing posture I'd till then happened to observe only in city and rural police commanders." A. P. Chekhov, *Ostrov Sakhalin (iz putevykh zapisok)* (Vladivostok: Izdatel'stvo Rubezh, 2010), 144.

"Your Worship, please order me unfettered, my legs are getting sick," one of them begged.

"No, dear, you'd be able to escape, so you're going to wear the chains. 'Wait a while,' that's what the court would tell you."

"But how many days were you at large?" Semënov asked.

"I was captured on the sixth."

"A minor offense," Semënov said, as if to himself.[30]

"But gentlemen, this here is a nobleman, a landowner from V–sk Province," Sh. pointed out for us.

Before us stood a short, 30-year-old man with a red beard, very embarrassed and blushing. He obviously still retained something good in himself and was ashamed of his present condition.

"Your Worship, please move me from here," he quietly requested.

"No, brother, stay here. You shouldn't have done your crime and you should be a landowner, but you tried to escape from here," Sh. superciliously answered him.

"But that was really long ago and now I've changed, I'm rehabilitated."

"Wait, brother, you'll stay," and Sh. left for the exit.

In the next ward a quite typical conversation unfolded.

"Your Worship," a penal laborer appealed to Semënov, "what decision have you reached on my petition?"

"Your petition?"

"Yes, I gave you a petition a long time ago."

"I didn't get anything; but whom did you send it through?"

"I gave it to the gentleman prison commander to pass on to you."

A dumbshow: The penal laborer and the comrade prosecutor look questioningly at the prison commander, who stands smoking with the most careless look on his face, gazing at his cigarette's bluish smoke. We watch, waiting for what he will say.

"Did you receive a petition from him?" Semënov finally asked.

". . . Yes, I received it."

"And why didn't you pass it on to me?"

"Don't you see? I looked at it, it proved to be complete rubbish, and so as not to trouble you with reading rubbish, I destroyed it."

---

30. Punishments for escape were based on the number of days a prisoner was at large. Most escapes were really intended by prisoners as brief respites from labor before they would voluntarily turn themselves in. Semënov is saying that the punishment Shel'king ordered for this prisoner was too stiff given the short time he was missing.

"I'm very grateful for your generosity," answered Semënov, barely control-
ling himself. "But I humbly request this not be repeated. Nothing good will
result if we snatch documents from each other!"

We left the ward.

"Do you want to see the aid station?"

"With pleasure," and we entered there.

But what was this? Was this truly an aid station? The sour stench of sheepskin hit the nose, and across the floor were spread unfinished sheepskin jackets, pieces of leather, and wool. Of the 10 people in the room, some were
conspicuously ill. Sh. saw what an impression the scene was making on us.

"We have such little space in this prison that I moved the furriers here;
there're also four patients here, but the ward is bright and not crowded, as you
see."

We expressed interest in what their illnesses were, and it turned out that two
had pulmonary tuberculosis. And here these tubercular wretches had to breathe
in dust, small hairs from the wool, and the sour smell of dank sheepskin.

"Is the doctor often here?" I asked a patient.

"No, he rarely shows," came the answer.

Very rarely, obviously, for otherwise, how could it be possible to allow for
a furrier shop in a sick ward?[31] (I also heard from the penal laborers certain
opinions about the local prison doctor, but since they weren't proven to me, I
won't convey them.)

At that moment Doctor D. was rather smilingly speaking with a consump-
tive penal laborer. I approached him.

"Here, allow me to introduce this patient to you. He's a specialist in the
operation of mechanisms," D. suggested.

"No, Your Worship, that's not true."

At that very instant we were leaving the aid station.

"What kind of mechanism does he work on?" I asked.

"You really don't understand the local terminology? A 'mechanism' is a
throat, and there are many cases of strangulation behind him. Here, they say—
'working on a mechanism.'"

My skin froze. Could this really be joked about?

Yes, great habituation is necessary, and not only the penal laborers but the
educated here are completely used to the horror and disgust that need be
endured.

---

31. I've since received in Petersburg a letter from the prison doctor, in which he writes that he
told the prison commander about this. [Akif'ëv's note.]

"Gentlemen, the penal laborers are now at work, so I've nothing to show you. Actually, let's go see Elena Bubelis."

We were led to a separate ward where we saw a woman in a shawl, with hair tumbling from beneath the shawl and purely masculine gestures and a masculine voice. Seménov began speaking with her, and I noticed she frequently mistakenly referred to herself as a man. In general, she was very strange.

"Who is this?" I quietly asked Dr. D.

"This is a renowned hermaphrodite," he answered, "and she already has several murders to her credit. To the extent needed, she was a man or a woman, and later committed her crime as 'Elena Bubelis.'"

I took her photograph, and we left the ward.

"Why is she kept separately?" I asked.

"She really can't be held with the men *or* the women. She carries on like the most depraved of women with the men, but with the women, like a man. Moreover, we don't have a women's section in Korsakovsk Post. All the women go into cohabitation, and those not chosen are sent to Aleksandrovsk."

From there we went to the isolation cells. These were small, tight kennels, some completely dark, since the only light entered through a small aperture in the door. There was no stool, table, or cot in a cell. In general, getting whisked away to the *prigioni* [prisons] in the Venetian doge's court would have been somewhat preferable. Here was that tool by which means Mr. Sh. wanted prisoners to submit.

All the cells were occupied. Above each was a placard on which the prisoner's offense and length of sentence were spelled out. Written on each was: "By order of the prison commander."

We all turned as one to notice a man in a black topcoat and freeman's clothes, for the most part. Upon our appearance, he rose from the bare floor on which he'd been lying and screwed up his eyes, because we'd opened the door and the light was too bright for him.

"Who's this?" Seménov asked.

"But this is a peasant-formerly-exiled.[32] He was getting drunk and making a row in the street, so I stuck him here."

The punishment seemed too strict to me. A prison commander seizes a freeman, beyond his jurisdiction, because he was getting drunk, and sticks him in a dark isolation cell of the chains prison.

---

32. After penal laborers completed their labor terms, they transferred to exile-settler status for ten years. Following completion of this stage, they entered the peasant estate (*krest'iane*). This theoretically allowed them to leave Sakhalin for the Siberian mainland, but many lacked the resources to do so.

Having completed our tour of the chains prison, we went to the free prison. First we passed through the almshouse, where there were a lot of people. Many were old men.

"Just look at what old men we have, for example, this Turkoman," Sh. turned to us, grabbing a tall, gray, stern, but very handsome old man by the beard. Obviously, Mr. Sh. doesn't consider holding and tugging a man by the beard—moreover, an old and wretched man—to be vile. What's there to say about this?

From the almshouse, we went to the joinery and saw the furniture there.

"This is the furniture they assemble for me," Mr. Sh. announced to us.

Rumor has it they actually do work exclusively for the local administration here.

We finished our prison tour inside of barracks once more. These ones were outside the prison walls; they were military barracks the command had courteously assigned to the prison. All the barracks were completely full, everyone was in them per order of the prison commander, and some had been there quite a long time. We listened to them complaining from all sides to Mr. Sh.

In one barracks it was quite cold, and when I mentioned this Sh. said: "It's nothing, it's rather warm."

Meanwhile, there could clearly be seen among the prisoners' graffiti that covered the stove from top to bottom—"The barracks is very cold." Again, I drew everyone's attention to this. And when an inmate, for his part, complained to Seménov that he felt extremely cold, the prison commander ordered the stove heated up.

In one barracks there sat a cannibal. This wretch had fled prison with a comrade. They couldn't find anything to eat, were starving terribly, and each nearing death. However, this one was much stronger than his comrade. So, having lost his mind, he murdered his comrade and cut a chunk of flesh from his side and cooked it. Satiating his hunger, he put the rest of his terrible food in a sack and continued on. But he was captured that same day and the meat found inside his sack. He was in an isolation cell because he seemed not to have been tried yet, and the penal laborers shun and condemn him.

It's said that in Aleksandrovsk Prison there's a certain cannibal, or "us-eater," as the penal laborers say, who's tried human flesh several times and found it tasty. I have his photograph, and to me, he appears psychologically abnormal.[33]

Having walked around the barracks we were just about to leave. Suddenly, there was a shout: "Mister Comrade Prosecutor, I ask that you listen to me!"

---

33. This is another reference to Lombrosianism, but also to degeneration theory, which was a forerunner of eugenics.

"This is Gorskii, who I was telling you about," Mr. Sh. explained. "He's here for evading work."

The door had been ordered unlocked and we were getting ready to leave. The young man stood in front of us, terribly excited: he was shaking all over as if in a fever, his face flushed then going pale, feverishly bloodshot eyes gleaming. Normally, he should sooner have been interned in a hospital and not in an isolation cell at all.

"Listen to what they're doing to me!" he practically shouted. "The irons were injuring my legs, I couldn't work, and I told them about this; they stuck me in the isolator; when I said that wasn't fair, the guards beat me; I complained to the prison commander, but he ordered me dragged to the shed and they gave me 29 birch blows. Is this really possible without a court? Mister Comrade Prosecutor, I'm asking you, look into my case! I can't take anymore."

"Very well, calm down, I'll look into it and see whether you're right," answered Semënov, and we headed toward the exit; but he rushed to catch up to us:

"Do we really not have rights, laws? If you don't help me, I'll complain to the Prosecutorial Office! This is so impossible . . ."

I was stunned. "So impossible!" was throbbing inside my head, droning in my ears and throughout my entire being. Sooner from this accursed place. Hardly bidding my companions goodbye, I practically ran like an escapee to the steamer.

That evening there was a social occasion. All Korsakovsk Post's intelligentsia were there. While conversing I was unable to obfuscate and not say that my impressions were horrible; this of course became common knowledge within a minute. During the farewells the assistant district commander, a university man, told me:

"I hear you've gotten a very poor impression. You were here all of several hours and, strictly speaking, saw very little, so your impression may be unbalanced. It's a pity you can't stay here longer, for in any case, you could then unpack your impressions and might reach a more equitable opinion."

Perhaps this is so, but no thank you. I've seen quite enough: those scenes won't fade from memory for the rest of my life. They were all too horrible, too burdensome. Sooner away, as possible, to get further from this terrible island, where thousands of living people are decomposing, physically and morally.

CHAPTER 10

From Sakhalin to Vladivostok

Typhoon—The steamer *Sungari*—Vladivostok—Miagkov leaves without a passport

8/21 October. Sea of Japan. We're sailing from Vladivostok to Japan aboard the steamer *Tartar*, which belongs to something called the Canada Company, and recalling our past impressions..

On Sakhalin, we returned from Korsakovsk Post to Mauka,[1] on the Tatar Strait, but we didn't stop there and continued directly south to Vladivostok. Our sailing was comfortable, if one disregards the typhoon we suffered on the 27th: the steamer was bobbing like a woodchip and the howling and noise were terrible; but the sight of our nonplussed Captain E. calmed everyone down. By midday, we were passing, he said, through the center of the typhoon. The scene was very strange and impressive; up to that moment the waves, however enormous, had been foaming and moving in one direction and there was no fog, just patchy gray clouds dropping low, low. But we suddenly ended up in a fog, and the way we'd been rocking changed; the water boiled as if inside a kettle; the waves chaotically rushed one after another in various directions, combined to tower like two fighting horses on their hind legs, and, having risen

---

1. The original text reads "Мацка" (*Matska*), which is almost certainly a typographical error that has been corrected here.

like an entire mountain, smashed into foam and then fell headlong downward, forming a deep hollow; the wind blew in fits to the right, left, and from all sides. There was such tossing, you couldn't keep your feet, and the whole ship was creaking. But then the wind began blowing more evenly, the pounding stopped, the waves went straight, in a single direction, and the fog dispersed.

The captain entered the dining room.

"Congratulations, gentlemen; the barometer's rising, and in several hours it'll be calm."

"But was this a serious storm, Captain?"

"Well, it was nothing to me, and fortunately we're out of it, though I don't especially like this rocking and fogginess."

After several hours the wind did indeed relent, and next day the weather was marvelous, and everyone crawled on deck. Traveling with us is an elderly lady with her young daughter, who is completely her little girl and is the wife of an officer who's been dispatched to the war. She was so seasick she could not recover, and was lying in a chair on deck.

"Please look at this lady," her mother jokingly complained. "I was completely seasick yesterday, lying down, terribly afraid, and not moving. But she clambers into my berth. 'Mama,' she shouts, 'let me in with you, I'm scared.' It was uncomfortable, but if you please, this grown-up lady stretched herself out nonetheless."

We continued to get acquainted a bit, and by evening were eating and telling jokes with each other.

The steamer *Sungari* left much to be desired.[2] First, the filth and stench on deck were impossible. Kitchen scraps were tossed straight into a sluice on the deck; but the sluice was all stopped up and reeking. We complained to the captain; he ordered it cleaned and said:

"Gentlemen, you can't do a thing with this bunch of Chinamen. They're extremely filthy and lazy. Every run, I change the restaurateur and the staff, but one's no better than the others."

Indeed, they all serve you reluctantly, you can request nothing, and there's practically nothing in the buffet. The only sound is them banging crockery.

In general, Chinese Eastern Railway steamers don't recommend themselves. For instance, aboard the steamer *Ninguta*, if I'm not mistaken, there was the following incident. The engineer didn't exactly know how much coal they had when the steamer went to sea. There soon proved to be not enough coal, and they had to burn benches and other wood items just to get themselves to shore.

---

2. This is the ship that brought them from Sakhalin to Vladivostok.

We reached Vladivostok at noon on the 29th. How good it was to leave the sea.

Encircled by wide-ranging green mountains, Golden Horn Bay embraces a collection of various sorts of ships. Along the shorelines there's an entire forest of the masts of Chinese junks, visible among them is the gigantic black corpus of a Volunteer Fleet steamer with its monstrous smokestacks, and in the middle of the bar, warships white as swans, reflected in the water's calm, blue surface, are visible. The thick band of the city stretches along the bay's western shore; its varicolored buildings run from the hilltops down to the bay's waters and continue for several versts along the coast, to the end of the bay.

Whereas Vladivostok is nice from the sea, this can't be said when you find yourself on its streets. It bumps up against the mountains' steep slopes so that walking is rather fatiguing. The seemingly beautiful buildings on the sea prove not to be so close up; the streets are unpaved and covered with wood, only Svetlanskaia has been paved, and that just a couple years ago. I remember it being done two years ago here. The traffic along Svetlanskaia Street—it being the primary as well as the shopping thoroughfare—is always heavy. The dust would get kicked up, and the wind generated entire clouds of this dust; it got under your shirt, filled your mouth and nose, got in your eyes, and in a moment coated you all over; within five minutes, it was like you'd been colored gray. Now, Svetlanskaia isn't dusty. Over the past two years many new buildings have been built; that the city is prospering and improving is commonly noticed.

The city has a huge deficiency—this is its lack of good hotels. During my previous visit I stayed in the Pacific Ocean Hotel, but for lack of anything better wound up in such a damp, smelly, dark room that I've ever since recalled it with a shudder. The prices were very dear. Having this time chosen not to stay at the Pacific Ocean, we stayed in the Efimovs' rooms, but we didn't come out ahead. They gave us a loathsomely filthy kennel for three rubles a day. First thing, we ordered them to remove the carpets, in which there was more dust than material; but we had to reconcile ourselves to the dirty wallpaper and the old, worn-out floor. All the prices were steep: drinking a couple of coffees came to a ruble or more; several spoons of jam with your tea cost a fifty-piece.

We decided not to eat amid such grime and preferred to go to the Shuina restaurant, on Aleutskaia Street, where, though it wasn't cheap, we ate well. Nearly every evening, we found ourselves there with company hailing from Sakhalin. There was little to do, we just had to sit and wait for the steamer, and the boredom was terrible; we had only the one above-mentioned family home, so we simply had to spend evenings at the restaurant. There's no real entertainment in the city. There was something akin to the Pacific Ocean Hotel; we went there once; onstage was some singing, dancing, and declaiming, but it was all so repulsive, so vulgar, and so far from art that we preferred to devote our evenings to "peaceful gustatory delight," as one of our company put it.

Throughout the city it was completely unnoticeable that a war was going on. Groups of Chinese were walking the streets, some Chinese were building a carriageway on Svetlanskaia Street, and Chinese were trading in the shops; in a word, relations apparently haven't changed, and judging by life in Vladivostok, no one would say the Chinese are our enemies. In general, the war wasn't intruding on the city or anything, save for only the crowded restaurants and the unusual sight of a lot of military doctors wandering the city. They've been so quickly brought to the city in such a crowd they can't find anything to do.

Not long ago, there arrived the sailors who defended the Peking Legation during the 50-day siege there; they dashingly strolled about the city, their St. George's ribbons decorating their chests.[3]

Karl Ivanovich finally arrived with the rest of the travelers from the expedition, but they didn't stay long and the next day left for Japan aboard the steamer *Dafne*, leaving Nikolai with us.

We'd decided to leave aboard the Volunteer Fleet steamer *Kherson*, which was supposed to depart soon. The steamer is enormous and very nice, and we dreamed about traveling in it pleasurefully. We were stopped as we were buying our tickets the day before its scheduled departure. The war ministry had requisitioned the *Kherson* for an indeterminate time, and the money for our tickets was returned. It was a terrible disappointment, for while we were awaiting the *Kherson* all the steamers left in front of our noses, and now we had to sit and wait. We ran to the agency, but there did not turn out to be a steamer soon. We were going to have to wait several days. To our good fortune, the *Tartar* arrived, and we didn't waste time buying tickets.

There was a mishap before leaving Vladivostok. Miagkov lost the international passport I'd obtained for him the day before.

"Your passports, gentlemen!" we were asked.

Miagkov thrust his hand inside his pocket, and his faced suddenly expressed not just fear but bewilderment.

"I must have lost my passport!" he said.

"Please, look hard for it!"

---

3. Boxers and the official Chinese army together besieged the Russian and other foreign legations from June to August 1900.

Miagkov ran to the cabin, rummaged about, and returned.

"It really isn't anywhere and must have fallen out in the city!"

"Then you'll have to stay, since I can't let you leave without a passport."

It was a rotten situation; there were several minutes before departure and it was too late to obtain a passport, and our tickets had already been taken and our baggage was on the steamer.

"Here's my certificate from the Academy of Sciences; in this extreme case it seems this can substitute for a passport," said Miagkov.

"Show me, please! No, unfortunately this can't substitute for a passport."

"What can I do?"

"You'll have to wait, get a new passport if you don't find the lost one, and leave on the next steamer."

"This is really inconvenient for me!" Miagkov erupted frightfully.

"Contrarily, perhaps you have a local official here who might personally confirm your identity?"

Fortunately there was A., a member of the court, who certified that he knew Miagkov, and the officer stamped the academy identification and Miagkov was able to go in peace. The officer laid down his papers and left, and the steamer, having whistled three times, slowly departed the wharf and crept toward the causeway. We've assumed that until Odessa itself, we'll avoid any unpleasantries over the passport.

A light wind is blowing and it's choppy today. The phosphorescence is remarkable. The whole sea seems to be on fire.

Chapter 11

Japan

Moji, and the railway to Nagasaki—The Hotel Nagasaki—Browsing the shops—A reception for the Crown Prince—A stroll to Mogi—In the teahouse—We depart—Something about the Japanese

9/22 October. Here came Japan. The steamer, cutting swiftly through the water, sailed between picturesque green islands; everyone was on deck; the air was fresh and clean; a light wind was already carrying the scent of vegetation to us; islands covered in greenery of the most varied hues rose from the sea's emerald waters.

Terraces have been formed at the bases of the islands, and there are even more on their mountains' gently inclining slopes. The Japanese value the little bit of land they have; not a single bit is left free; the work of human hands and hard labor are visible everywhere, as it must not have been easy to shape from out of the twisting mountain slopes—overgrown with forests—the terraces as well as the irrigation that leads everywhere, nor to prepare the soil for so refined a cultivation as that of rice. But the climate here comes to the Japanese's aid. It's late October, but it's as warm as a nice spring day. Life bubbles over. Wherever you look, the entire sea is covered with little barges and boats whose white sails seem, from a distance, to be the wings of gulls.

Moji was already visible ahead.[1] A string of buildings appeared on the flat shoreline, beyond was an entire forest of masts, topped by factory chimneys, puffs of smoke reaching into the clear blue sky, and beyond them, silhouetted on the horizon through the delicate haze of the morning fog, a ring of mountains.

We'd just managed to breakfast when the steamer dropped anchor. An entire horde of Japanese officials immediately came to us aboard three cutters: tiny, sunburnt, and ugly, they clambered through the cabins, wrote stuff down, inspected each passenger personally, and again noted something. This took a whole hour. Finally, the passengers began to settle into the boats; but this was not at all comfortable: we crowded down one and the same ladder into the cutters and sloops, and there also came a barge with coal loaders, both Japanese men *and* women; nonetheless, following these torments, we made some sense; found a boat and got to shore; stopped at the customs house at the river's mouth; our baggage was moved there for a thorough inspection and they rummaged through everything, though nothing was chosen for a duty.

During this there was a slight misunderstanding: Miagkov had put several dozen already exposed photographic plates in a basket, and an official absolutely wanted to uncover them to see what they showed; Miagkov understandably did not want him to expose them to the light and began telling him so with great effort.[2] It was a good thing we had our academic certifications, for they were read and we were allowed to go on, otherwise all that work could have been lost; but whereas we had to pay nothing to customs, we did for some reason have to pay 4 yen (4 rubles) to the boatmen. The train station proved to be nearby, and we now got our tickets and handed over our baggage. It was something staggeringly cheap: for a 10-hour trip in first class, we paid 3 r. 10 k., and not a kopek for baggage; our group was traveling as a foursome, as we [Akif′ëv, Miagkov, and Nikolai] had been joined by a young artillery officer who was very ill; he could eat almost nothing, was terribly thin, and had lost his voice completely; he was traveling to Japan for medical treatment, and asked us to be his companions.

Having snacked in a small Japanese eatery, we settled into a car. The cars are quite narrow; there are leather divans along the walls for sitting; the windows are large and electric lights are in the ceiling.

The train started off. Outside the window additional strange scenes flashed before our eyes. There were descending terraces of yellow rice paddies between hills covered in dark green pine trees.

---

1. Moji is located on the northern tip of Kyushu Island.
2. Early photographic plates were sensitive to ultraviolet light.

Rice grows only with an abundance of water; therefore, to plant it, the surface must be made perfectly horizontal and in such a way as to hold water. This is achieved through the construction of low dikes between which the expanse is smoothed out, and there is an entire system of small, primitive sluices through which the necessary amount of water flows. Before harvesting, the water is drained and a thin crust forms, under which a layer of sticky silt is found. So much work is needed to clear the soil! And the Japanese perform all this industriously.

What marvelous highways you see here, what picturesque bridges, and there are swarms of people everywhere, in the fields and on the roads; here before my eyes is a village: I glimpse clean, little domiciles with beautiful, concave roofs; beside each is a garden with fruit trees.

Life and labor are bubbling everywhere. Your eyes don't get fatigued as in Germany by uniform plowed fields and uninterrupted gridlines, straight as arrows. No! Here scenes change by the minute, one better than the next. The various hues are staggering, the soft, delicate hues harmonizing uniquely with each other; there are the bright green bamboo groves, the palm tree with its beautiful pinnate leaves, the dark-green pine groves, the orange trees and such, bending beneath the gravity of bright red fruit, and the mountains, plains, and ravines with rings of transparent snow, through which pass bridges and smooth, flat roads beautified by their twisting and turning here and there, and there is the emerald sea, and above all this, the necessarily blue sky and the warm sunrays pouring down.

Two Japanese in their winged outfits entered the car. The clothes on them were silk, on their heads were bowler hats, in their hands were umbrellas and necessities, and on their feet was the invariable wooden sandal with two transverse supports [the *geta*]. When they sat down on the divan and their costumes' flaps happened to open, nothing proved to be underneath. This combination of a bowler and umbrellas with an absence of underwear is very characteristic of the Japanese. They immediately took out small wooden boxes, like those we put dried fruit in in Russia, and having opened them began to eat, working swiftly with little sticks in place of knife and fork. This interested me very much and I was intrigued by what was there; I saw rice in one but some sort of varicolored tidbits in the other. I was inspired to try some.

At a station we bought these boxes and decided to taste for ourselves. Though everything was so tidy and pretty in appearance, we were quite disappointed. The rice proved to be completely flavorless, though those tidbits were something surprising. What wasn't there! There were pieces of crab, mushrooms, sweetbread, meat, some sort of vegetables, and even something completely unrecognizable to us. Everything was without salt, had the odor of

Japanese soy sauce, and was strongly seasoned with pepper. The entire breakfast cost 15 cents. I decided to experiment with myself as to whether a European
stomach could endure this Japanese cooking, so I ate the entire breakfast—i.e.,
both boxes. It proved to be nothing—I survived. All this was downed with weak
tea, which is served together with cups and teapot here for 3 cents.

Evening came. The electricity was turned on in the car; it got boring, and
everyone started dozing off. From behind the little barrier separating us from
2nd class (our car was *mixte*[3]), talk and laughter were drifting over. We went
there; a lot of people, both Europeans and Japanese, were there. They were
eating the same Japanese breakfast and laughing. We joined them, and at the
next station bought some beer and began treating ourselves.

I ended up sitting next to a young Japanese woman. There were several of
them, and they were sitting cross-legged on the divans. For something to do,
I offered my neighbor a beer. She accepted and drank like it was nothing, and
then for her part poured a cup of tea and offered it to me, while Miagkov
was treating another. Seeing we'd already gotten familiar with the Japanese
women, the group became even more jovial. Jokes and barbs rained down.
After a little while the Japanese women, visibly tired, turned to the wall and
wanted to sleep, but lying down was not at all possible because it would spoil
their intricate coiffures. Therefore, they assumed the most improbable poses:
one was sitting with her legs on the divan, leaning her forehead on the wall,
another was somehow completely bent over and propping her forehead on a
sac-voyage. Had I bent over like that I would have died on the spot.

But then the train stopped: Nagasaki. We exited the car: we decided to leave
our baggage at the station and go to a hotel by ourselves. I'd stayed at the Hotel
Nagasaki two years ago, and this was the only hotel I knew. We hired rickshaw pullers and the small barouches they carry people in, and headed there.
Trotting at the quick with their varicolored lanterns and shouting from time
to time, the rickshaw pullers brought us to the city. It's completely dark at night
in the narrow streets, and you're amazed at how quickly and smartly the pullers dart through them, not once bumping into anything. To the right opened
the bay, on whose dark surface ships' lights were burning like stars, and across
from it everything was lit up by the Hotel Nagasaki's glimmering lights. We'd
traveled for a good half hour. Having finally arrived, we entered.

"Are there rooms available?"

"There are, if you please!"

"What's the cost?"

"Full room and board is 8 yen per person per day."

---

3. "Mixte" is French for "mixed."

Expensive, but what could we do? Finding a hotel at night is inconvenient. We consented. We were taken to the third floor and settled in our room. Miagkov and I got one large room illuminated by a pair of small lamps. There was one bed, albeit such that two could sleep in it. Nevertheless, 16 rubles a day for a room with one bed, even with meals, is expensive.

Our traveling-companion officer terribly wanted a drink.

"Give us something hot, like tea," he said.

"No," they answered, "it's already late and there's nothing hot."

"Well, just give us beer."

They brought it. A half bottle of beer was 60 cents, and ten foul cigarettes were 30 kopeks. In a word, we'll need to leave tomorrow.

10/23 October. We awoke and tried the coffee. They were offering the Devil-knows-what sort of slops. We changed clothes, breakfasted downstairs, and went out to the street.

Mornings in Nagasaki are so nice. The shipping lane remains covered in morning fog. An entire city of ships of all flags and colors is visible through the blue haze. Cutters and sloops glide in all directions along the green water. Along the shore, behind and in front of planted trees, run the rickshaw pullers in their tight-fitting black uniforms and white hats resembling mushrooms, each tethered to his small two-wheeled barouche. A coolness comes in from the water.

We browsed the shops all day. There was so much of interest and originality that we didn't want to leave the street; so the money drained from our pockets. On this first day Aleksandr Gennad'ich spent 70 rubles on purchases. You see all the wares as you walk along the street and without entering the shops because all the stores remove their exterior walls.

Certain branches of Japanese fine arts are much superior to the European. Woodcrafts, for example. All these little tables, cabinets, and boxes have been fashioned with such taste and care; there are very fine and beautiful wooden mosaics, and everything costs a copper. You're simply amazed by how they can labor so cheaply. And the products of tortoise shell and elephant tusk are of strictly high artistry here. The execution is so richly designed and so charming. Were you to see the Japanese woven materials that are in geishas' clothes and girdles, you'd be ashamed to look at our tastelessly clumsy calico prints. We have in Europe nothing similar to such rich patterns, so unique and delicately beautiful. Indeed, the Japanese are in general superior to Europeans in many ways. Foremost is that you see in every little item the craftsman's love and interest in his work, and not a mechanical or routine attitude toward the business. Take the enameled divider (the cloisonné). Could we really produce this? In one shop, I spotted our Russian prints among the Japanese. I was simply embarrassed by our mediocrity.

Returning to our lodgings, we encountered Hirasa, a translator I know who was with me during my first visit to Nagasaki, and with his help found a different hotel—the Japan Hotel, very modest but completely satisfactory.

We decided to relocate immediately after lunch, because remaining in the Hotel Nagasaki was impossible. When we were given our bill, it turned out we owed 40 rubles even though a full day hadn't elapsed.

We came over to the Japan Hotel this evening, where the rooms are smaller, there aren't even bells, and the dining room is two small rooms. It's a two-story wooden building surrounded by trees. Full board and lodging is 3 dollars a day. Moreover, the staff here speak Russian, and many Russians live here. You hear Russian in the dining room. A door of our room opens onto a balcony and below us are some sort of tropical plants. It's peaceful and quiet.

16/29 October. This morning we went to see a reception for the Japanese heir.[4] White flags with a red ball in the middle adorned the entire city, and in the lane, the ships were showing off all possible colored flags. There was a lot of traffic in the streets; the crowd thickened closer to the train station. Students and teachers from city and village schools were lining the streets practically up to the very entrance to the city. The boys were in gray Japanese khalats with standardized kepis on their heads, and the girls were in the most brightly colored uniforms.

We alighted from the carriage without entering the station and stood behind a row of girls. Instantly, a policeman approached and asked the interpreter something.

"He is asking who you are. I told him you are Russian doctors."

But this didn't end the matter, and the policeman approached again and took a visitor's card from the interpreter. In general, the police, in their clean uniforms and white gloves, with worried expressions on their faces, stroll through the streets showing great diligence. Having tired of standing, several members of the public, Aleksandr Gennad'ich among them, moved off to the side and were sitting on some rocks without disturbing anyone. But this rather displeased the policeman and he ordered everyone to stand and move closer to the crowd. Hirasa explained it this way:

"Before, according to Japanese custom, no one could stand during a visit from the Mikado or the princes, but now, according to the new custom, no one can sit and everyone must stand."

---

4. "The Japanese heir" refers to Yoshihito Shinnō, who in 1912 would become Emperor Taishō.

The day was hot and humid, and the prince didn't arrive for quite a long time. It got boring and rather tiresome. I was sustained only by the thought that the children had been standing in place since seven o'clock in the morning, whereas we were hardly able to stand for an hour.

For something to do, we began talking to the schoolgirls and took to watching passersby. There was much of curiosity in these Japanese, festively dressed for such a gala event. Top hats were on some, and when their khalats billowed from the ground up, their naked legs could be seen, and on their feet those same wooden sandals were clopping. In Japan, you're struck by this generally odd combination of the Japanese with the European. Here came a Japanese dressed like a European, in yellow shoes even, yet beside him was a Japanese woman with blackened teeth. The custom of women blackening their teeth is ancient but still retained, even though it is very ugly. Women blacken their teeth after marriage to show they don't want to please other men. Along with the top hats, canes, and gloves goes the custom of eating with a pair of chopsticks, and, even more so, a universal bathhouse for men *and* women.

In the evening we went to see the Osuwa and Dai Baksu temples, and for us, of course, several questions arose concerning Japanese religions. We turned to our translator for explanations. But no matter what we got out of him, he couldn't answer a single question. I've really never encountered such complete ignorance regarding religious matters. Take the most backward peasant from some backwater place, take a Korean, and you'll see they have their own—sometimes very childish and illogical—religious views. Here, you encounter such indifference to these same issues that you're simply astounded. This isn't just my opinion: a local Orthodox missionary we became acquainted with says the very same thing. One day, I'd bought an entire collection of Japanese idols (it's said they have up to 70 here). Each of them was named, but the vendor universally called them all Buddhas. This is what Buddhism has become here!

And so we were standing and waiting for the prince. Finally, there was movement in the crowd and several rickshaw pullers appeared. Everybody took off their hats. Walking ahead in formation were various military personages, and behind them, a young, thin man, looking very typically Japanese, who was exchanging bows with people and raising his hat slightly. Coming behind him were several dozen Japanese in European top hats and clothes, but at times in something that even Raspliuev would be embarrassed to dress.[5] The crowd greeted the prince rather silently. There were no cheers or ovations and then, with the same silence, it dispersed.

---

5. In Aleksandr Sukhovo-Kobylin's 1854 comedy *Krechinskii's Wedding* (*Svad'ba Krechinskago*), the character Raspliuev is both a swindler and a scoundrel.

I had several unexpected encounters today. The city of Nagasaki is such that you never know who you'll meet here. The Romans used to say: "All roads lead to Rome"; now it's possible to say with regard to the East: "All roads lead to Nagasaki." Traveling along the street I met a comrade, the odd-duck Doctor K., and we were overjoyed.

"Greetings!" he shouted. "Where are you coming from and where are you going?"

"Yes, indeed, I'm coming from distant wanderings and trying to get home to Russia. And you?"

"I arrived from Odessa aboard the hospital ship *Tsaritsa*, and we're leaving today for Port Arthur.[6] Pay us a visit if you can."

After lunch, Miagkov and I were aboard the *Tsaritsa*. We met the doctors. It turns out K. has been traveling a lot. He served in Central Asia; and here it was three years later that we were running into each other in Nagasaki. Theirs was a large group—8 doctors and 40 Sisters of Mercy, who, by the way, aroused such interest on the streets that foreigners would stop and stare at them.[7] Unfortunately, we didn't have long to speak. The second whistle blew and, having said good-bye, we left.

This evening, in the Russian hospital, we met with Dr. Korsakov, who weathered the Peking siege, took ill, and is now getting treatment here.[8] We'd not seen him for a very long time, so we blabbed all night with pleasure.

Living in our hotel is another one of those victims from China. This young man is the railway foreman Iasinskii. He was wounded, captured by the Chinese, and put in prison, where he was beaten, starved, and held for two months. He himself can't understand why he was set free and not killed. All his comrades in the prison were executed, and several times he was ordered executed but spared and remanded to prison again. He informed us that the engineer Verkhovskii was killed and that his head has been recovered.[9]

28 October/10 November. We've finally finished our shopping. Tired of knocking about the shops and wanting to go for a spin outside the city, Miagkov,

---

6. Port Arthur was a Russian military post on China's Liaodong Peninsula. Doctor K. would have been one of the doctors assigned to care for Russian troops fighting in Manchuria and Peking.

7. The Sisters of Mercy were a religious community of nurses. See John Shelton Curtiss, "Russian Sisters of Mercy in the Crimea, 1854–55," *Slavic Review* 25, no. 1 (1966): 84–100; and Laurie Stoff, *Russia's Sisters of Mercy and the Great War: More than Binding Men's Wounds* (Lawrence: University Press of Kansas, 2015).

8. This hospital was apparently connected to the Russian navy's earlier use of Nagasaki as a coal depot.

9. Both Europeans and Chinese beheaded each other during the years leading up to and during the Boxer Uprising.

Nikolai, and I went with our interpreter to Mogi, a little spot situated behind a small pass, not far from Nagasaki.

The road from Nagasaki to Mogi is strikingly beautiful, and smooth and even; it's a strangely twisting, wide ribbon that passes through mountain slopes above green paddies, gardens, and houses, and leads to a valley, disappears inside a bamboo grove, and then follows the bank of a stream, whose thin, regimented bamboos—their pale-green pinnate leaves swaying in the breeze—reflect on the cold, limpid water. The surrounding mountains are clothed in a luxuriant, variegated covering. Light green patches of bamboo, and smooth, glimmering palm leaves, can be seen amid bright green pines, and there rise dark hills with summits of gigantic camphor trees and thuya, which in places bend beneath the weight of their bright red fruit. The forest's aromatic scent is carried on the fresh mountain breeze. Mountain streams burble over rocks, and from above, from out of a delicate, blue sky, the sun pours its warming rays upon the entirety of this marvelous scene. Here lies Mogi—a fishing village stretching along the shore of a small bay that is encircled by green mountains.

The Mogi Hotel, so frequented by tourists, is here. But it no longer attracts the great crowds of foreigners it used to. There was no pretty Ochiya San, the hotel's friendly hostess. We were disappointed, and the whole scene around us somehow faded. Ochiya San, it turned out, has sold the hotel and lives in Inasa, across from Nagasaki. But I still have such fond memories of the days I spent there, two years ago. At her table we found instead the visitor log where we'd signed our names. It's full of the most enthusiastic praises for Ochiya San's place, in all languages, especially Russian. Mogi still has all the same picturesque places, but without her there it doesn't measure up.

Having eaten several crabs and drunk beer, we returned to Nagasaki, having quickly decided to find Ochiya San; Miagkov and I went to Inasa after lunch, but had no luck there. We did find her mother, Oya San, who said that Ochiya had gone on a ride with her husband. With nothing to do, we sat and talked for a little while.

Oya San is a wealthy woman and owns several houses and a large property in Inasa. She wasn't feeling well living near the water and decided to build a home on a mountain, where the air is fresher and cleaner and there's a splendid view of the shipping lanes and the city. Reinforcing the property and building a terrace alone came to 11 thousand rubles. She's built on this terrace a small, spacious, neat house with a surrounding gallery, and has planted trees around the terrace. The extra rooms she lets out to lodgers.

"I love company," she said, "and though I don't need the money (I'm sufficiently wealthy), I like having people in my home. I have mostly Russians living here, there are a few foreigners, though I don't like them and can't understand their language."

It's nice to see how well certain people regard Russians here. She gets 3 rubles a day from Russians and those she doesn't consider foreigners, and 4 from the rest.

We left her place for our residence. Along the way, we learned we could play billiards in a small teahouse. This place, built when the Russian squadron was here, is a sort of officers' club. Photographs of our mariners cover the walls, and we even found there several we know. The Japanese serving woman knew all their names.

Having returned to Nagasaki, we went this evening to see the geishas, so as to complete a whirlwind tour of the city, so to speak. Inviting two other Russians, we went under Hirasa's leadership to a teahouse. They asked us upon entry to remove our shoes and gave us slippers. We gathered in a large room and sat on pillows on the floor. We were brought tea and beer. The geishas soon appeared. Upon entering, they bowed to the ground and then sat down to eat. The dances were not a triumph this time. They danced a lot, with and without masks, with fans and with shawls, to the sounds of a drum and long, three-stringed guitars [shamisens]. It was very original, and at times likely decent. But earlier, two years ago, I watched the geisha dances and they were truly *grazioso* [graceful], filled with a kind of singular charm. This time, the geishas were very young and obviously just learning to dance. It got rather monotonous and wasn't graceful at all.

While the geishas were dancing a little girl approached Aleksandr Gennad'ich and sat down beside him. He wrote something down in his notebook. She took his pencil and notebook and painstakingly began tracing out numbers, turning to him for approval. There ensued a very touching scene. The geishas were singing and dancing, but Aleksandr Gennad'ich was holding the little girl on his knees and watching her diligently produce her naive scribbles.

29 October/11 November. In the morning the *Sydney*, a French company steamer, arrived and we repaired to it.

Once again, there was an adventure with Miagkov. He'd gone into the city for some reason, and was returning by himself from there to the ship. The ferryman didn't understand a word of Russian, and Aleksandr Gennad'ich knew the same amount of Japanese, and he was completely unable to explain which steamer he had to go to. The boatman took him through the entire road, but time was running out. The second whistle had already sounded. We were starting to get worried whether Miagkov would arrive in time, for if he didn't hurry, it was going to be a big disappointment: all our things were already on the steamer, and we had absolutely no money to spare. Finally, with five minutes to go, he showed up, terribly agitated and cursing the ferryman. The steamer weighed anchor and slowly proceeded towards the exit from the bay.

Farewell Japan, farewell, you wonderful, unique country!

Having briefly familiarized myself with her, I have a rather strange impression. For natural beauty and mild climate, not many places on Earth can compare, and not in vain do many Russians travel from Vladivostok to Japan as if to a dacha or to get treatment. The roads are splendid. It'd be nice if we had something similar to Japan's country roads. The railways are reasonably priced and outstandingly maintained. Japan's military and commercial fleets are not inferior to Europe's. Japan's literacy may be similarly praised: it's hard to find an illiterate Japanese man or woman. There are a bunch of schools in the cities and villages. Aspects of life demonstrate Japan's high culture. But somehow persisting alongside this are the wooden chopsticks, the oddly discomfiting wooden sandals, the beautiful coiffures that prevent women from sleeping, women's blackened teeth, and along with this their rather casual morality. If, with their often pretty faces and absolute femininity, Japanese women make a pleasing impression through a kind of unique feline grace, Japanese men are straightforwardly unpleasant. They are short and usually have ugly faces. Their manners are sugary and unctuous; they talk as if choking with pleasure, crouching and rubbing their knees, though this doesn't prevent them from being terribly cruel and vindictive. I was witness to such a fact.

It was two years ago in Korea, in the city of Chemulpo.[10] My comrades and I were getting ready to leave for Russia and had summoned the Korean porters to get our baggage. There was a lot of baggage, and a whole crowd of Korean coolies had gathered in front of the hotel entrance. One of them, wanting to get work, entered the hotel. None of us were in the lobby at the time. The hotel owner began pushing the porter out and ordering him to wait at the entrance to be summoned.[11] The Korean didn't understand and wouldn't leave, so the owner and his son then threw the Korean out by his neck, and in the meantime hit him with an iron poker in the head so that he fell down bleeding in the street.

I was coming onto the porch at just that moment. Seeing me, the Koreans shouted and dragged me to the wounded man, whose head was streaming

---

10. Today, Chemulpo is called Incheon.

11. Akif'ëv fails to mention here that the owner was Japanese, perhaps assuming his readers would already realize this.

with blood. I shouted for the interpreter, and through him the crowd explained what had happened. The Koreans complained to me about the Japanese owner and said this wasn't the first incident, that in general the Japanese were terribly cruel toward them. And this in Korea!

Incensed by this, I sent the interpreter for the police, and a comrade asked to meet with the British consul, since there was no Russian consul. A policeman arrived, and I saw that he, too, was Japanese.

"What happened here?" he asked.

"This owner here struck and injured a Korean, so a charge needs to be written up."

"But why does this matter to you?"

"How can it not matter, when I see that a man has nearly been murdered?"

"This is an affair between Japanese and Koreans here, but you're a Russian, so you have no right to interfere."

The British consul similarly recommended the matter be left alone if we didn't want trouble.

Another similar instance happened with us in Korea, in the city of Qingamno.[12] Having completed our Korea expedition, we'd transferred our baggage to the steamer. Everyone was onshore, but unbeknownst to us, one of our laborers had gotten drunk and was still in the city. After a little while, he came to the shore, covered in wounds. The palm of his hand had been sliced in half, his forehead was covered with gashes, blood was oozing from his shoulder through a cut in his sheepskin jacket. Bathed in blood, he explained he had gotten into a row in the street and broken a window in a Japanese home. Suddenly, a whole swarm of Japanese fell upon him with knives and would probably have killed him had he not been so big and strong. This is a vignette of their morality. Yet, what can be demanded of an immature people, who not so long ago were savage and uncivilized, when even worse atrocities occur in other countries boasting their own ancient cultures?[13]

So be it, however, but you feel rather sad leaving Japan; you stand on deck, unable to tear your eyes away from those magical islands fading little by little into the blue distance.

---

12. "Цинамно [Tsinamno]." What modern-day city this refers to is unknown.
13. This may be a veiled reference to the pogroms in Russia.

Afterword

As the 1900 Chukotka Expedition was undergoing its death throes half a world away, Tsar Nicholas II asked V. M. Vonliarliarskii for a personal report on its progress. Based on his knowledge at the time (which was probably nonexistent, given the absence of communications), Vonliarliarskii told the emperor that rumors of American prospectors on the peninsula were fictitious and that his Russian expeditionaries were having "no misunderstandings whatsoever" with either their American counterparts or the local inhabitants. But then Vonliarliarskii was made privy to a telegram that had been forwarded to the Mining Department on 13 September 1900. Bogdanovich had earlier complained to Russia's San Francisco consulate, which in turn telegraphed his misgivings to the Mining Department. Bogdanovich accused the Americans of ruining the expedition; the Nome military command of terrorizing him and his team; and Vonliarliarskii of turning the expedition over to the (in his words) "hands of piratical stockbrokers." He condemned Vonliarliarskii for making him hire the Americans in the first place and even suggested that Vonliarliarskii might have manipulated the selling of shares in the East Siberian Syndicate. Bogdanovich moreover complained that the military servicemen St. Petersburg insisted be attached to the expedition simply granted the Americans an excuse to scuttle it. The Mining Department called Vonliarliarskii in to answer these charges. He denied them, of course, and was cheeky enough to suggest that Bogdanovich be warned about the "tactlessness" of his communiqué.[1]

Sometime during that following winter Vonliarliarskii summoned Frederick Baker, his English business partner, to St. Petersburg and asked Bogdanovich, who had since returned from the Pacific, to join them. Vonliarliarskii claims in his account that he wanted to sort through the expedition's problems with both men. Following their meeting, he concluded that Bogdanovich's lack of English, his nervous wife's presence aboard ship, and Vonliarliarskii's own refusal to give up his rights in the East Siberian Syndicate had laid the bases for both misunderstandings between the expeditionaries and Bogdanovich's suspicions. All the same, observed Vonliarliarskii, despite the problems and lost workdays, the expedition did manage to investigate Cape Dezhnëv as well as the bays Koliuchinsk, Mechigmen, St. Lawrence, and Providence. Gold had been found, albeit in limited quantities. In his account, Vonliarliarskii adds that he brought charges in an American court against Captain Jahnsen for the purpose of redeeming Bogdanovich's honor. It is not clear when he did so, but this comports both with Bogdanovich having communicated with the San Francisco consulate and the geologist's later claim of encountering the Alaska customs agent Joseph Evans in that city. Vonliarliarskii writes that Jahnsen was found guilty of both disobeying and depriving Bogdanovich of command over the journey. But it is unclear whether he suffered any punishment. In any case, Vonliarliarskii adds that he and Bogdanovich were forced to bring the charges themselves, without help from their government; and in general, he portrays St. Petersburg's role as unhelpful to the Chukotka Expedition.[2]

Immediately following the expedition, Vonliarliarskii went into business with John Rosene, a Seattle entrepreneur. In 1902 they formed the Northeast Siberian Association, which took over the East Siberian Syndicate's Chukotka concession. This company's activities have been detailed by other historians, but it is worth mentioning here that it failed to turn a profit and contributed to "the anti-American uproar that occurred in Russia in 1906, immediately after numerous violations of its charter by Americans became public." Thomas C. Owen adds that it won "the dubious distinction of being Russia's most vilified corporation."[3]

---

1. V. M. Vonliarliarskii, *Chukotskii poluostrov: ekspeditsii V. M. Vonliarliarskogo i otkrytie novogo zolotonoskogo rainona, bliz ust'ia p. Anadyria, 1900–1912 gg.* (St. Petersburg: Tipo-litografiia K. I. Lingarda, 1913), 21–23.

2. Vonliarliarskii, *Chukotskii poluostrov*, 23–24; and K. I. Bogdanovich, *Ocherki Chukotskago poluostrova* (St. Petersburg: Tipografiia A.S. Suvorin, 1901), 79.

3. Thomas C. Owen, "Chukchi Gold: American Enterprise and Russian Xenophobia in the Northeastern Siberia Company," *Pacific History Review* 77, no. 1 (2008): 50. See also Nikolay Ivanovich Kulik and Anastasiya Alekseevna Yarzutkina, "Gold of Chukotka and Foreign Investments: Institutional

Vonliarliarskii's reputation as a mendacious profligate never recovered. During the mid-1920s he emigrated to Germany, where he continued to write his self-serving and factually questionable memoirs. Owen writes that nothing more is known about him, except that he died there in 1939.[4] However, a Russian-language Wikipedia entry states that in 1924, while still in Russia, Vonliarliarskii was arrested and briefly imprisoned before being allowed to emigrate the following year; that he lived in Germany but also in France; and that he died in 1946, in Karlsbad.[5]

Vonliarliarskii published his version of the 1900 Chukotka Expedition in 1913, whereas Bogdanovich published his much earlier, in 1901. It will be recalled that when Bogdanovich and the other Russians were preparing to board the *Iakut* and depart the peninsula, Jahnsen left first and sailed the *Samoa* straight into a storm, with Evans onboard. Bogdanovich writes that when he met Evans in San Francisco, the latter told him he had petitioned California officials to strip Jahnsen of his captain's certification for this and other reasons. It is not known whether Bogdanovich truly reconciled with Vonliarliarskii, as the latter implies. But the geologist's wounds were certainly still raw in 1901: "It is difficult to put into words all the bitterness, sufferings, and strenuous efforts there were for me with the *Samoa*. With what lying, deception, and impudence surrounded me, instead of an energy and willingness for collective labor that could have, under different circumstances, yielded outstanding scientific and practical results."[6] Having been prevented from carrying out a full investigation, Bogdanovich claimed he was unable to draw conclusions about Chukotka's gold reserves. He argues that the unfortunate decision to hire Jahnsen, Roberts, and the other Americans absolved him of any responsibility for the expedition's failure.[7]

Unlike Vonliarliarskii, Bogdanovich's association with the 1900 expedition did not limit his prospects. He continued as a state geologist, and in 1901 was awarded the Geographical Society's Konstantin Medal. He served as director of Russia's Geological Committee from 1914 to 1917. Bogdanovich left Soviet Russia for his native Poland, and in 1921 secured a professorship at the Krakow Mining Academy. He died in Warsaw in 1947, having spent the last nine years of his life as director of the State Geological Institute.[8]

Approach," *Middle-East Journal of Scientific Research* 15, no. 3 (2013): 410–11; and John J. Stephan, *The Russian Far East: A History* (Stanford, CA: Stanford University Press, 1994), 88.

　　4. Owen, "Chukchi Gold," 79–80. Vonliarliarskii's third and final volume of his memoirs is: V. M. Vonliarliarskii, *Moi vospominaniia, 1852–1939 gg.* (Berlin: Russkoe natsional'noe izdatel'stvo, 1939).

　　5. Вонлярлярский, Владимир Михайлович — Википедия (wikipedia.org), accessed 3 May 2022. The source provides no citations for this information.

　　6. Bogdanovich, *Ocherki*, 80.

　　7. Bogdanovich, *Ocherki*, xi–xii.

　　8. Богданович, Карл Иванович — Википедия (wikipedia.org), accessed 3 May 2022.

One of the only non-Russian expeditionaries whose afterlife can be traced is the *Samoa* captain Edward Jahnsen. His career continued many more years, indicating that Evans's and Vonliarliarskii's actions against him had little effect. Jahnsen was born in Norway on 25 September 1856, and at some point emigrated to the United States, where he met and married another Norwegian immigrant. They had two sons, one of whom, Oscar, pursued a law enforcement career. In 1970 Oscar was interviewed for the Earl Warren Oral History Project. He recalled his father being a shipbuilder as well as a captain. "I couldn't have asked for a better father or a better mother," he told interviewers. His father was a man who worked hard and lived hard. "[Y]ou might say there were wooden ships and iron men. [Laughter] My father came up the hard way. Most of the men were heavy drinkers. A lot of them sailed the good ship 'Rock and Rye.'" Jahnsen apparently never owned his own boat, but he was indeed a thirty-second-degree Mason. After captaining the *Samoa* he captained the *San Pedro* for several years off the California coast. In September 1901, after expressing regret at President McKinley's assassination, he reportedly had to defend himself against an "anarchist." Three years later, he began working for the McCormick Lumber Company, of St. Helens, Oregon, overseeing construction of its vessels and commanding at least one on its maiden voyage. In 1905, he lost his captain's license for thirty days after sailing the steamer *Cascade* full speed through the fog and running it aground. The *San Francisco Call* newspaper nonetheless praised Jahnsen as a "veteran skipper and one of the most efficient and best liked mariners in the Pacific coast service." Jahnsen had a stroke in 1914, but a news item the following year shows him transporting passengers between San Francisco, San Diego, and Hilo, Hawai'i, at cut rates. While delivering pilings for the Pearl Harbor drydock that same year, he suffered a second stroke. His third and final stroke happened while he was hanging out with friends during a late night in the pilot's office, at the Wilson Brothers Shipyard in Astoria, Oregon. He was moved to a hospital in his hometown of Oakland, California, but never regained consciousness. Five days later, on 29 August 1916, with his family at his bedside, he died. The *St. Helens Mist* eulogized: "No master mariner on the Pacific coast was more highly esteemed among friends and acquaintances than Edward Jahnsen." He had a reputation, it added, for being careful. Jahnsen retained memories of the Chukotka Expedition until his death, telling this same newspaper about it shortly beforehand. The obituary promised the newspaper would soon publish his oral memoir, though it does not seem to have done so.[9]

---

9. Oscar J. Jahnsen, "Enforcing the Law against Gambling, Bootlegging, Graft, Fraud, and Subversion, 1922–1942," interview by Alice King and Miriam Feingold Stein, May–July 1970, Regional

As for the hapless Aleksandr Gennad'evich Miagkov, following his adventure to the Far North he managed to live a long life. Though never completing his engineering degree, he was named second-in-command of a gold-mining expedition to Abyssinia (modern-day Ethiopia) in 1904. He married Vera V. Savinkova, a sister of Boris V. Savinkov, one of the top leaders of the Socialist Revolutionaries' Combat Organization—Russia's fiercest terrorist group prior to the Soviet period. In 1918 Miagkov moved to Zhytomyr, Ukraine, and then later to Poland. His family eventually settled in Prague. He belonged to the Uspenskii Brotherhood, an Orthodox sect the Soviets deemed a "White émigré organisation" and ordered abolished in 1951. Miagkov died in Prague, in either 1952 or 1960.[10]

Finally, let us return to our diarist and the author of *To the Far North*. During the Russo-Japanese War (1904–5), and despite his affection for Japan, his friendships there, and his reported misgivings about the conflict and the Russian army's abilities, Akif'ëv served with a Cossack detachment in Korea. On 7 December 1905, with the war over and Akif'ëv on his way home, he found himself dining with fellow officers in a restaurant in Irkutsk. Perhaps they were drinking their sorrows away or celebrating the end of hostilities, or maybe they were just being reckless, but somehow a loaded pistol fell from one of their pockets and, when it hit the floor, discharged a bullet into Akif'ëv's stomach. He was rushed to a surgeon who removed the bullet, but three days later, on 10 December, at the age of thirty-three, he died of sepsis.[11] So ended Akif'ëv's brief, adventurous life, a life partially preserved by the diary translated here.

Oral History Office, Earl Warren Oral History Project, 77/139 G, Bancroft Library, University of California, Berkeley, 3, 6; "Mariner Knocks Anarchist Down," *San Francisco Call*, 27 September 1901, 11; "Brief City News," *San Francisco Call*, 18 April 1905, 6; "Lumber Carrier on Maiden Trip," *San Francisco Call*, 13 January 1913, 15; "Shipping Notes of the Week," *St. Helens Mist*, 9 October 1914, 1; "Commodore of McCormick Fleet Passes Away," *St. Helens Mist*, 1 September 1916, 1, 5; and [Advertisement], *Honolulu Star-Bulletin*, 31 July 1915, 32.

10. Мягков Александр Геннадьевич (narod.ru), accessed 4 May 2022; Jan Holzer and Miroslav Mareš, *Czech Security Dilemma: Russia as a Friend or Enemy?* (Cham: Palgrave Macmillan, 2020), 157.

11. V. V. Korsakov, "Bezvremenno pogibshaia sila," *Russkie vedomosti*, 23 April 1906, 4. The anonymous author of a website on Akif'ëv speculates that his killing may have been intentional. The author bases this on the rather flimsy evidence that Akif'ëv had mailed a "basket" worth of correspondence home to his sister but she never received it. The author suggests Akif'ëv may have known secrets the government wanted kept hidden. Based on the materials available, there is no way to evaluate this claim. Cf. "Bezvremenno pogibshaia sila," *Livejournal* "Безвременно погибшая сила": nlr_spb—LiveJournal, accessed 29 October 2021.

Acknowledgments

I first came across Akif'ëv's *To the Far North* while researching my book on the Sakhalin penal colony. But it was not until 2020, when my own transglobal plans were thrown into disarray by the COVID-19 pandemic, that I decided to translate it for publication. Finding myself homeless and dependent on others, I had some basis for identifying with Akif'ëv. Then, after managing to secure more or less suitable living arrangements, I, like others, watched on TV as it was happening the 6 January 2021 assault on the Capitol. This was followed a year later by Russia's invasion of Ukraine. My work on the translation slowly continuing, these events led me to see in Akif'ëv's diary a relic of modernism and its travails, a relic that reveals the influence that nationalism, imperialism, and racialism apparently had on a young Russian doctor's developing *Weltanschauung* and his interactions with non-Russians. It brought me no great comfort that many of the preconceptions and prejudices Akif'ëv suffered from remain with us today, precisely due to the lingering presence of nationalism, imperialism, and racialism. But I also found cause for hope, given the humanitarian ideals Akif'ëv also expresses. Human beings will undoubtedly remain works in progress, but his diary suggests that travel and interaction with others can serve as a major avenue of self-improvement, one that over time allows the better angels of our natures to supersede our demons.

I want to acknowledge my editor Amy Farranto, as well as Christine Worobec, my anonymous readers, and my copyeditors for their enthusiastic support and assistance in helping me realize this project. Their comments and criticisms, along with those of my known reviewers Lee Farrow and Elena Campbell, helped me revise the manuscript for the better. Thanks also to Ellen Labbate and everyone else at NIU/Cornell University Press, to Ted Weeks for his professional encouragement, and to Rob Ebersol for creating the book's informative and quite necessary map. I also wish to acknowledge the librarians and archivists at the Library of Congress, Russian National Library, and El'tsin Presidential Library for making some of the sources used here, including most importantly Akif'ëv's photographs and his book itself, available for online access. Finally, thanks to Dinah Loculan, whose support for this and my other projects has been steadfast, and to whom this translation is dedicated.

10 August 2023
Panglao, Bohol
Philippines

SELECTED BIBLIOGRAPHY

Akif'ëv, A. *Na dalekii sever: iz dnevnika krugosvetnogo puteshestvennika*. St. Petersburg: Tipo-litografiia E. Tile preemn., 1904.

Allen, Douglas W. "Information Sharing during the Klondike Gold Rush." *Journal of Economic History* 67, no. 4 (2007): 944–67.

Allen, Robert V. *Russia Looks at America: The View to 1917*. Washington, DC: Library of Congress, 1988.

Anderson, Benedict. *Imagined Communities: Reflections on the Origins and Spread of Nationalism*. New York: Verso, 1991.

Behringer, Paul J. Welch. "Images of Empire." *Russian History* 45, no. 4 (2018): 279–318.

Belich, James. *Replenishing the Earth: The Settler Revolution and the Rise of the Anglo-World, 1783–1939*. Oxford: Oxford University Press, 2009.

Betts, Raymond F. *The False Dawn: European Imperialism in the Nineteenth Century*. Minneapolis: University of Minnesota Press, 1975.

Bogdanovich, K. I. *Ocherki Chukotskago poluostrova*. St. Petersburg: Tipografiia A. S. Suvorin, 1901.

Bogoras, Waldemar. "The Chukchi of Northeastern Asia." *American Anthropologist* 3, no. 1 (1901): 80–108.

Brooks, Jeffrey. "Official Xenophobia and Popular Cosmopolitanism in Early Soviet Russia." *American Historical Review* 97 (1992): 1431–48.

Bruner, Jane Woodworth. "The Czar's Concessionaires: The East Siberian Syndicate of London; A History of Russian Treachery and Brutality." *Overland Monthly* 44 (1904): 411–22.

Burbank, Jane, and Frederick Cooper. *Empires in World History: Power and the Politics of Difference*. Princeton, NJ: Princeton University Press, 2010.

Burbank, Jane, and Mark von Hagen, eds. *Russian Empire: Space, People, Power, 1700–1930*. Bloomington: Indiana University Press, 2007.

Burns, Adam. *American Imperialism: The Territorial Expansion of the United States, 1783–2013*. Edinburgh: Edinburgh University Press, 2017.

Butlin, Robin A. *Geographies of Empire: European Empires and Colonies c. 1880–1960*. New York: Cambridge University Press, 2009.

Carlson, Leland H. "Nome: From Mining Camp to Civilized Community." *Pacific Northwest Quarterly* 38, no. 3 (1947): 233–42.

Carlson, T. H. "The Discovery of Gold at Nome, Alaska." *Pacific Historical Review* 15, no. 3 (1946): 259–78.

Chekhov, Anton. *Sakhalin Island*. Translated by Brian Reeve. London: Alma Classics, 2019.

Connelly, John. *From Peoples into Nations: A History of Eastern Europe.* Princeton, NJ: Princeton University Press, 2020.

Cravez, Pamela. *The Biggest Damned Hat: Tales from Alaska's Territorial Lawyers and Judges.* Fairbanks: University of Alaska Press, 2017.

Crisp, Olga. "Russian Financial Policy and the Gold Standard at the End of the Nineteenth Century." *Economic History Review* 6, no. 2 (1953): 156–72.

Crisp, Olga. "The Russo-Chinese Bank: An Episode in Franco-Russian Relations." *Slavonic and East European Review* 52, no. 127 (1974): 197–212.

Crist, David S. "Russia's Far Eastern Policy in the Making." *Journal of Modern History* 14, no. 3 (1942): 317–41.

Cruikshank, Julie. "Images of Society in Klondike Gold Rush Narratives: Skookum Jim and the Discovery of Gold." *Ethnohistory* 39, no. 1 (1992): 20–41.

[Dalberg-Acton, John]. *Home and Foreign Review* 1 (July 1862): 1–25.

Demuth, Bathsheba. *Floating Coast: An Environmental History of the Bering Strait.* New York: W. W. Norton, 2019.

Demuth, Bathsheba. "Geology, Labor, and the Nome Gold Rush." In Mountford and Tuffnell, *Global History of Gold Rushes,* 252–72.

Drummond, Ian M. "The Russian Gold Standard, 1897–1914." *Journal of Economic History* 36, no. 3 (1976): 663–88.

Ducker, James H. "Gold Rushers North: A Census Study of the Yukon and Alaskan Gold Rushes, 1896–1900." *Pacific Northwest Quarterly* 85, no. 3 (1994): 82–92.

Eklund, Erik. "Creating a Global Industry? Geology, Capital, and Company Formation on the Goldfields of the Industrial Age." In Mountford and Tuffnell, *Global History of Gold Rushes,* 184–205.

Etherington, Norman. "Reconsidering Theories of Imperialism." *History and Theory* 21, no. 1 (1982): 1–36.

Gentes, Andrew A., trans. and ed. *Eight Years on Sakhalin: A Political Prisoner's Memoir.* New York: Anthem Press, 2022.

Gentes, Andrew A., trans. and ed. *Russia's Penal Colony in the Far East: A Translation of Vlas Doroshevich's "Sakhalin."* New York: Anthem Press, 2009.

Gentes, Andrew A. *Russia's Sakhalin Penal Colony, 1849–1917: Imperialism and Exile.* New York: Routledge, 2021.

Helms, Andrea R. C., and Mary Childers Mangusso. "The Nome Gold Conspiracy." *Pacific Northwest Quarterly* 73, no. 1 (1982): 10–19.

Hobsbawm, E. J. *Nations and Nationalism since 1780: Programme, Myth, Reality.* Cambridge: Cambridge University Press, 1992.

Kaspe, Sviatoslav. "Imperial Political Culture and Modernization in the Second Half of the Nineteenth Century," translated by Jane Burbank. In Burbank and von Hagen, *Russian Empire,* 455–93.

Korsakov, V. V. "Bezvremenno pogibshaia sila." *Russkie vedomosti,* 23 April 1906, 4.

Kramer, Lloyd S. *Nationalism in Europe and America: Politics, Cultures, and Identities since 1775.* Chapel Hill: University of North Carolina Press, 2011.

Kulik, Nikolay Ivanovich, and Anastasiya Alekseevna Yarzutkina. "Gold of Chukotka and Foreign Investments: Institutional Approach." *Middle-East Journal of Scientific Research* 15, no. 3 (2013): 408–19.

Lukoianov, I. V. *"Ne ostat' ot derzhav . . .": Rossiia na Dal'nem Vostoke v kontse XIX–nachale XX vv.* St. Petersburg: Nestor-Istoriia, 2008.

Marks, Steven G. *Road to Power: The Trans-Siberian Railroad and the Colonization of Asian Russia, 1850–1917*. Ithaca, NY: Cornell University Press, 1991.

McGhee, Robert. *The Last Imaginary Place: A Human History of the Arctic World*. Chicago: University of Chicago Press, 2007.

McKee, Lanier. *The Land of Nome*. New York: Grafton Press, 1902.

Miagkov, A. "V poiskakh za zolotom." *Russkoe bogatstvo* 8 (1901): 102–59.

Mogilner, Marina. "Beyond, against, and with Ethnography: Physical Anthropology as a Science of Russian Modernity." In *An Empire of Others: Creating Ethnographic Knowledge in Imperial Russia and the USSR*, edited by Roland Cvetkovski and Alexis Hofmeister, 81–120. New York: Central European University Press, 2014.

Mogilner, Marina. "Russian Physical Anthropology of the Nineteenth–Early Twentieth Centuries: Imperial Race, Colonial Other, Degenerate Types, and the Russian Racial Body." In *Empire Speaks Out: Languages of Rationalization and Self-Description in the Russian Empire*, edited by Ilya Gerasimov et al., 155–89. Boston: Brill, 2009.

Mountford, Benjamin, and Stephen Tuffnell, eds. *A Global History of Gold Rushes*. Berkeley: University of California Press, 2018.

Owen, Thomas C. "Chukchi Gold: American Enterprise and Russian Xenophobia in the Northeastern Siberia Company." *Pacific History Review* 77, no. 1: 49–85.

Owen, Thomas C. *Russian Corporate Capitalism from Peter the Great to Perestroika*. New York: Oxford University Press, 1995.

Remnev, A. V. *Rossiia Dal'nego Vostoka: Imperskaia geografiia vlasti XIX–nachala XX vekov*. Omsk: OmGU, 2004.

Remnev, Anatolyi. "Siberia and the Russian Far East in the Imperial Geography of Power." In Burbank and von Hagen, *Russian Empire*, 425–54.

Renner, Andreas. "Defining a Russian Nation: Mikhail Katkov and the 'Invention' of National Politics." *Slavonic and East European Review* 81, no. 4 (2003): 659–82.

Rogger, Hans. "America in the Russian Mind: Or Russian Discoveries of America." *Pacific Historical Review* 47, no 1 (1978): 27–51.

Rytkheu, Yuri. *A Dream in Polar Fog*. Translated by Ilona Yazhin Chavasse. New York: Archipelago Books, 2005.

Schimmelpenninck van der Oye, David. *Toward the Rising Sun: Russian Ideologies of Empire and the Path to War with Japan*. Dekalb: Northern Illinois University Press, 2001.

"Skorbnyi listok." *Voenno-meditsinskii zhurnal* 216 (June 1906): 396.

Slezkine, Yuri. *Arctic Mirrors: Russia and the Small Peoples of the North*. Ithaca, NY: Cornell University Press, 1994.

Stalker, Nancy K. *Japan: History and Culture from Classical to Cool*. Berkeley: University of California Press, 2018.

Stephan, John J. *The Russian Far East: A History*. Stanford, CA: Stanford University Press, 1994.

Vanderlip, Washington B., and Homer B. Hulbert. *In Search of a Siberian Klondike*. New York: Century Co., 1903.

Vaskevich, P. "Ocherk byta Iapontsev v Priamurskom krae." *Izvestiia Vostochnago Instituta* 15, no. 1 (1906): 1–30.

Vaskevich, P. *Ocherk byta Iapontsev v Priamurskom krae*. Verkhneudinsk: Tipografiia A. D. Reifovicha, 1905.

Vitte, Sergei Iu. *Vospominaniia*. 2 vols. Moscow: Izdatel'stvo sotsial'no-ekonomicheskoi literatury, 1960.

Vonliarliarskii, V. M. *Chukotskii poluostrov: ekspeditsii V. M. Vonliarliarskogo i otkrytie novogo zolotonoskogo rainona, bliz ust'ia p. Anadyria, 1900–1912 gg.* St. Petersburg:Tipo-litografii K. I. Lingarda, 1913.

Zabytaia okraina: rezul'taty dvukh ekspeditsii na Chukotskii poluostrov, snariazhennykh v 1900–1901 gg. V. M. Vonliarliarskim, v sviazi s proektom vodvoreniia zlotopromylennosti na etoi okraine. St. Petersburg: Tipografii S. Suvorina, 1902.

Ziker, John P. *Peoples of the Tundra: Northern Siberians in the Post-Communist Transition.* Long Grove, IL: Waveland Press, 2002.

Printed in the USA
CPSIA information can be obtained
at www.ICGtesting.com
LVHW091155010924
789842LV00005B/517

9 781501 774614